WHAT in the World is MELANOMA?

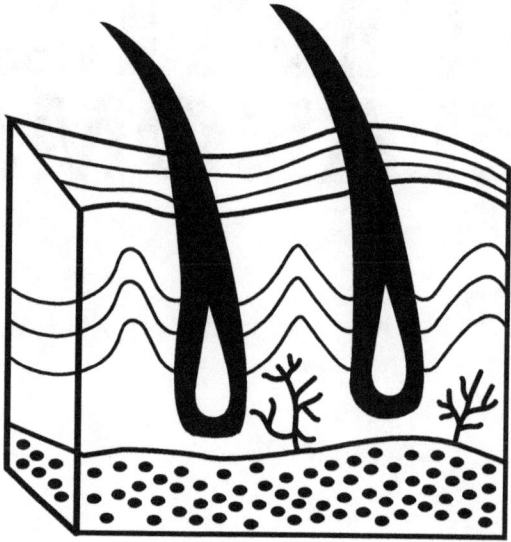

WRITTEN AND EDITED BY

Austin Mardon, Mohammed Ismael, Madiha Ansari, Adrienne Lam, Maria Ashraf, Aaliyah Mulla, Jannat Irfan, Hafsa Binte Younus, Abeer Ansari, Jennifer Pham, Angelina Lam, Paige Breedon, Massa Mohamed Ali

WHAT in the World is MELANOMA?

GOLDEN METEORITE PRESS

First Printing: 2021

Typeset and Cover Design by Brianna Purai

ISBN: 978-1-77369-255-5

What in the World is Melanoma?
Golden Meteorite Press
103 11919 82 St NW
Edmonton, AB T5B 2W3
www.goldenmeteoritepress.com

GM

PRESS

Table of Contents

History of Melanoma

Madiha Ansari

Cancer is one of the most dangerous and life-threatening diseases in this world; there is an abnormal and uncontrollable growth of cells in one's body, which makes cancer the second most reason behind someone's death, globally (World Health Organization: WHO, 2019). Melanoma is a type of skin cancer caused by the tumour of melanocytes; cells that produce melanin (Melanoma Treatment (PDQ®)–Health Professional Version, 2021). Since these cells divide and spread uncontrollably, it causes dark spots to be developed on one's body. Along with the arising cases, there have been many deaths which are caused because of melanoma; leading up to an estimate of 7,180 deaths in the United States in 2021 alone (Melanoma Treatment (PDQ®)–Health Professional Version, 2021). There are many factors that are involved in the development and spread of melanoma; high levels of sun exposure, genetic, and environmental factors (Berwick et al., 2015). Sun exposure is one of the highest risk factors for melanoma since there is no set amount

1

of time period in which the skin can get affected. In essence, whether the exposure to the sun is chronic or intermittent, it would cause melanoma or affect it in a certain way (Berwick et al., 2015). Genetically, research suggests that if an individual has a family history of melanoma, it increases their chances of developing melanoma as well (Mitchell et al., 2020). Additionally, families in which genes are inherited in an autosomal dominant manner, allows one to get diagnosed with melanoma at an early age (Mitchell et al., 2020). Moreover, the inheritance is continued through generations, and sometimes, an individual is at risk of developing a type of cancer which is associated with melanoma (Mitchell et al., 2020). Apart from the sun exposure, other environmental factors also play a role in causing melanoma. Some of these factors include the chemical components which an individual is exposed to; whether it is through food, atmospheric situations, or occupational conditions (Berwick et al., 2015). Furthermore, there are four subtypes of primary cutaneous melanoma; Superficial Spreading Melanoma (SSM), Lentigo Maligna Melanoma (LMM), Nodular Melanoma (NM), and Acral Lentiginous Melanoma (ALM) (Mitchell et al., 2020). To add on, other subtypes of melanoma also include Desmoplastic Melanoma (DM), Mucosal Melanoma (MM), and Uveal Melanoma (UM) (Mitchell et al., 2020). These different subtypes of melanoma would be mentioned further in this chapter. Moreover, melanoma has the capability to affect different age groups. Generally, research suggests that elderly men are at the highest risk for developing melanoma, however, melanoma is the most common in individuals ranging from ages 25 - 29 years old and the second most common is in people that are aged 15 - 29 years old (Melanoma Treatment (PDQ®)–Health Professional Version, 2021). Exceptionally, individuals who have a history of melanoma in their family, are more likely to develop it at an early age (Mitchell et al., 2020).

The first case of melanoma was discovered during the 1800s when John Hunter, a Scottish surgeon, reported on an unusual mass on the jaw of a 35 year old man (Lee et al., 2013). In 1812, Guillaume Dupuytren and René Laennec were one of the first individuals to describe la melanose and publish it in the Faculté de Médecine in Paris (Lee et al., 2013). In 1820, William Norris published various cases of melanoma, in a review, in which he mentioned what factors cause the disease to occur, as well what features play a role in the identification of melanoma (Lee et al., 2013). Along with that, Samuel Cooper, one of the surgeons at the time, introduced several advantages of an early surgery for melanoma, in which the concept of preventing other diseases can be made possible (Lee et al., 2013). In 1892, Herbert Lumley Snow highlighted the necessity of anticipatory gland dissection which can be used as a treatment for patients that are dealing with melanoma (Lee et al., 2013). Similarly, in 1907, William Sampson Handley emphasized the benefits of excising melanoma, dissecting the lymph node, or amputating when it comes to severe cases (Lee et al., 2013). Therefore, ever since the first case of melanoma was reported in the 1800s up until 1950, there has been research on the causes, consequences, and prevention of melanoma. Many surgeons have contributed to understanding melanoma by figuring out the best and worst possible case scenario for this cancer. Throughout the years after 1950, there have been discoveries which display what affect melanoma has and what can be used to affect melanoma, whether it is in a positive or negative way. For instance, from 1950-1960, Paul Ehrlich presented the possibility of the immune system contributing in repressing the spread of carcinomas (Lee et al., 2013). This idea was used to observe how the immune system prevented the spread of melanoma; an individual who has a stronger immune system would have a decreased incidence of melanoma (Lee et al., 2013). During the 1980s and 1990s, the concept of immunotherapy and chemotherapies were introduced (Lee et al., 2013). However, researchers who were identifying possible ways to prevent or cure melanoma were indecisive as to which therapy is likely to be the most effective (Lee et al., 2013). In the end, it was concluded that the effectiveness of a certain type of therapy depends on an individual and the type of melanoma they have (Lee et al., 2013). Nonetheless, throughout these years, many discoveries were made which showcased how important it is to identify and classify a disease, and then find different ways to cure that disease. Hence, melanoma was not just a type of cancer that was discovered in the 1800s, it was also an opportunity for researchers to enhance their knowledge when it comes to different types of cancer.

As mentioned previously, as melanoma is a type of skin cancer, it has different subtypes to it as well. One of the subtypes of primary cutaneous melanoma is Superficial Spreading Melanoma (SSM) (Mitchell et al., 2020). SSM is a type of melanoma which is responsible for 70% of all the melanoma incidences and is most likely to occur in patients that are around 50 years old (Mitchell et al., 2020). SSM is caused due to intermittent sun exposure; the appearance of the mole is large, no regularity or symmetry, and often exhibits different colours such as black, pink, and blue (Mitchell et al., 2020). SSM can be affiliated with the mutation in the BRAF gene that causes the protein, which is regulating cell growth, to be altered (Mitchell et al., 2020). Another subtype of primary cutaneous melanoma is Lentigo Maligna Melanoma (LMM), which causes 4% - 10% of the melanoma incidences (Mitchell et al., 2020). LMM occurs mostly in old patients and is caused due to a chronic exposure to the sun (Mitchell et al., 2020). The lesion, when it comes to LMM, appears flat and light brown in colour (Mitchell et al., 2020). Unlike SSM, LMM is associated with KIT, which is a protein that can lead to cancer if mutated (Mitchell et al., 2020). Nodular Melanoma (NM) is another subtype of primary cutaneous melanoma that has a 15% of incidence for all melanomas (Mitchell et al., 2020). NM can be caused by any one of the factors; whether it is sun exposure or other environmental factors (Mitchell et al., 2020). While NM does not account for any mutations in a particular gene, it is classified by the lesion that appears to be enlarged and/or elevated and is often pink, blue, or black in colour (Mitchell et al., 2020). Acral Lentiginous

5

Melanoma (ALM) is one of the four subtypes of primary cutaneous melanoma, which has an incidence of 5% - 10% of all melanomas (Mitchell et al., 2020). Similar to LMM, ALM is also associated with mutations in KIT. Furthermore, ALM is not caused by exposure to the sun, instead, other factors play a role in its developments (Mitchell et al., 2020). The lesion caused with ALM is dark, flat, and has irregular borders; they are most commonly found under the nail bed or on the palms (Mitchell et al., 2020). Apart from these four subtypes of primary cutaneous melanoma, Desmoplastic Melanoma (DM) is one of the subtypes of melanoma; accounting for 1% of the melanoma incidences (Mitchell et al., 2020). The features of DM vary; it can be displayed as a firm papule or can be mistaken as a cyst, and it is most likely to affect elders (Mitchell et al., 2020). Nonetheless, DM has a high risk of recurring if a wide excision takes place (Mitchell et al., 2020). Therefore, each one of these subtypes have exhibited to be a cause of a mutation, sun exposure, or other factors. Hence, they are identified and classified based on their appearance and causes.

Throughout history, different preventative measures have been undertaken by researchers in order to limit the spread of melanoma. One of the solutions is to reduce the time spent outside; limiting sun exposure (Berwick et al., 2015). However, this solution does not consider the necessity of other benefits that are gained from sun exposure. Along with that, melanoma is not only caused by a chronic sun exposure, but it can also be developed by intermittent exposure to the sun. Hence, other solutions were introduced including the usage of sun-protection products (Berwick et al., 2015). These products would help in preventing the direct exposure to sun as well as limit the sunburns that many individuals go through (Berwick et al., 2015). Apart from this natural treatment, there has been an on-going research on the effectiveness of different therapies and surgeries which can help in curing the spread of melanoma. As mentioned previously, chemotherapy is one of the treatment methods which can kill the uncontrollable cells and stop them from growing (Rebecca et al., 2012). However, the side effects of chemotherapy included proliferation of other normal cells in the body as well, leading researchers to stop recommending chemotherapy to patients (Rebecca et al., 2012). Nonetheless, advancements in technology have been made in which different methods can be used alongside chemotherapy; this would be discussed later in the book. Other types of therapies include targeted therapy and immunotherapy. Targeted therapy is a method which uses drugs to inhibit the mutations in tumour cells (Faries & Curti, 2018). For instance, mutations in the BRAF gene can be repressed if the therapy is effective; preventing the spread of melanoma (Faries & Curti, 2018). Meanwhile, immunotherapy is a treatment method used to push the immune system into recognizing the spread of abnormal cells, and as a result, destroy the malignant cells that are growing rapidly (Faries & Curti, 2018). Therefore, a lot of research has gone into testing the effectiveness of different types of therapies and natural treatment methods. These concepts would be further mentioned later in the book.

Melanoma is one the most dangerous types of cancer. Apart from the factors that cause melanoma to occur, there are several consequences of what might happen to individuals who are diagnosed with melanoma. A research conducted in 2010 evaluated melanoma patients who survived 2 months after they were diagnosed (Bradford et al., 2010). The researchers included all melanoma patients from 1973 - 2006; from which it was found that 89,515 of the patients survived for at least 2 months after they were diagnosed with melanoma (Bradford et al., 2010). In this study, 53.4% of the patients were male while 46.6% of the patients were female (Bradford et al., 2010). Along with that, Superficial Spreading Melanoma (SSM) was the most common subtype of melanoma; 94.3% of the patients were treated with surgery whereas 2.1% of the patients were treated with radiation (Bradford et al., 2010). It was found that from the 89,515 patients, 10,857 or around 12.1% of the patients developed at least one of the subsequent primary cancers (Bradford et al., 2010). From these 12.1% of the patients, 3094 of them developed melanoma as a subsequent cancer (Bradford et al., 2010). Along with that, the risk of developing subsequent primary cancers increased by 28% (Bradford et al., 2010). Furthermore, it was found that after 20 years of being diagnosed with melanoma, the risk of subsequent primary melanomas remained elevated (Bradford et al., 2010). Nonetheless, melanoma patients had the highest risk within the first year after being diagnosed; considering the increased surveillance on the changes that occurred on the lesions (Bradford et al., 2010). Moreover, the families of these patients also developed a high risk of getting melanoma due to the genetic factors and inheritance of a mutated gene (Bradford et al., 2010). The analysis of this study was histologically reviewed by pathologists, making this study credible in understanding the consequences of melanoma (Bradford et al., 2010). This study is one of the examples that explain why melanoma has been a serious issue since time. Along with the inheritance of the mutated gene and the possibility of passing on this cancer to other individuals, melanoma is also developed by sun exposure, which is something an individual goes through daily. Furthermore, melanoma patients, who have already gone through the treatment, have a higher risk of developing another, or the same, type of cancer again, leading to more deaths which are caused by melanoma.

In recent years, the cases of melanoma have increased rapidly. Due to the direct exposure from the sun, other environmental and genetic factors, as well as the occupational situations that are faced by some individuals, the increase in melanoma patients was to be expected. However, in order to prevent this disease, many researchers have recommended individuals to use sun-protective products or reduce the time they spend outside. Along with that, immunotherapy and target therapy are two of the treatment methods that researchers have deemed reliable enough for melanoma patients. Historically, many discoveries in order to understand melanoma, as well as many inventions in order to prevent melanoma, have been made; allowing the present researchers to acknowledge the severity of this cancer. In conclusion, melanoma is a type of skin cancer which causes different coloured spots, depending on the type of melanoma, to appear on the skin. The abnormal growth of cells, whether it is due to a mutation caused by sun exposure, environmental factors, or genetic factors, causes an individual to consider the possibility of death; in the most serious cases. The historical existence of research has suggested preventative measures and possible therapies, still, the cases and deaths from melanoma occur to this day. However, undoubtedly, the investigation into different types of melanoma continues to advance further, which adds on to the research that was done during the past years. Henceforth, the present and future of melanoma's causes, consequences, and treatments, would be discussed further in this book.

The First Acknowledgement of Melanoma

Adrienne Lam

When we hear about cases of skin cancer, we think it is a recent ailment, however, cancer has affected many people for centuries. It is important to consider the broader carcinoma, or cancer. A disease that arises once abnormal cells mutate out of control, cancer has a lengthy history. Humans and animals alike have been found with this disease, ancient manuscripts, fossilized bone tumours, and human mummies provided as evidence. Although the source of this disease is debatable, oncologist Siddhartha Mukherjee added in his book The Emperor of All Maladies, "Civilization did not cause cancer, but by extending human life spans, civilization unveiled it." This chapter will cover the first acknowledgement of melanoma, but also the various other carcinomas to provide a thorough understanding of melanoma and cancer.

Cancer has increased recently due to the aging world population in the last half-century, as well as the risky behaviour and carcinogens we expose to ourselves from the environment and various products we consume. It is important to consider other animals in addition to humans as carcinoma may be present in all living organisms. The oldest credible evidence of cancer in mammals were in fossilized dinosaurs and human bones consisting of tumour masses (Faguet, 2014). However, significant evidence of cancer in dinosaurs was found in a study involving fluoroscopy that screened over 10,000 species of dinosaur. Evidence of tumours and other abnormalities of tissue were found using computerized tomography (CT) scan. Many species were tested, however, only the cretaceous hadrosaurs, or the duck-billed dinosaurs which lived more than 70 million years ago had benign tumours while 0.2% of the species tested had malignant metastatic cancer (Faguet, 2014). Metastatic cancer is when the original cancer mass breaks off and travels in one's bloodstream to other areas in the body which leads to grave outcomes.

Origin of the Word "Cancer"

"Cancer" originates from the word "karkinos" from Hippocrates (460-370 BC), a Greek physician considered the "Father of Medicine". He initially used the words karkinos and karkinoma to describe non-ulcer and ulcer forming tumours. Non-healing swelling was called karkinos, while karkinoma was implemented for non-healing cancer. Cancer was precedingly dubbed "karkinos", greek for crab, because of a tumor's resemblance to the animal.

The word was subsequently translated to "cancer", the latin term for crab, by Roman physician Celsus.

The First Recorded Case of Cancer

Cancer has been around since the appearance of man, seen by the paleontological findings of tumours in animals. Ancient Egypt was the oldest documented case of cancer, dating back to 3000 BC by Edwin Papyrus (Fayed, 2020). Other documented cases include Ebers Papyrus dating back 1500 BC, or over 2,700 years ago. Eight cases of tumors on the breast were recorded on papyrus, which could only be treated using palliative treatment (Papavramidou, Papavramidis, Demetriou, 2010). Palliative treatment for cancer today involves palliative chemotherapy and palliative radiation (Fayed, 2020). These treatments do not cure the disease, but rather provide a sense of relief for the symptoms. These cancers that were discovered include skin, uterus, stomach, and rectum cancers. This case was handled with a hot instrument called "the fire drill", destroying tissue with a method known as cauterization (Papavramidou N, Papavramidis T, Demetriou, 2010). Through evidence and inscriptions, we can see that the Ancient Egyptians understood the difference between malignant and benign tumours, and were surgically removed similarly to how they are removed today (Fayed, 2020). The Edwin Smith Papyrus, an ancient medical text, states, "There is no treatment." He also referred to cancer as a lethal disease and designated it with 'the curse of the gods'. This belief continued into the Hippocrates (460-

270 BC) which will be discussed soon (Fayed, 2020).

Another Early Cancer Case

South Africa, also known as the Cradle of Humankind holds the earliest known case of one of the world's most deadly diseases. Osteosarcoma, an aggressive cancer, was found in an approximately 1.7 million year old foot bone by researchers through 3-D imaging. This body part belonged to a human relative. The South African Journal of Science published this discovery, proposing that the triggers of cancer are embedded in the human evolutionary past, despite modern lifestyles having increased cancer cases. University of the Witwatersrand study co-author Edward Odes states, "You can opt for the paleo diet, you can have as clean a living environment as you want, but the capacity for these diseases is ancient, and it's within us regardless of what you do to yourselves,"

Evidence of Cancer

Edward Odes and his colleagues used a method called micro-CT imaging. There was no clue to the cause of death of the hominin, or their age. However, it was recognized that the osteosarcoma would have hindered the person's ability to travel on foot, and cause agony (Weiss, 2000). His research team reviewed the hominin bone both 2-D and 3-D, capturing the differences in density, and looking into different perspectives of the fragments. The team came to diagnose the foot bone with osteosarcoma, a disease affecting mainly adolescents, by studying the abnormal growth pattern of bone tissue. Odes declared that "We compared the images, it was bingo." (Weiss, 2000). Additionally, an older fossilized tumour was analyzed by Ode's team. The 1.98 million year old skeleton had a benign growth located in a vertebrae. This was discovered by National Geographic Explorer-in-Residence Lee Berger at Malapa. This was provided as extra evidence of malignant cancer. "A tumor is new growth of bone or tissue, where you have a sliding scale from benign to malignant," says paleoanthropologist Patrick S. Randolph-Quinney, one of the investigating scientists. "On the benign scale, there are mechanisms that keep these tumors in check, so they are self limiting, or they reach a certain size and they basically stay there. Whereas cancer is the extension of that growth process without the control mechanisms." Furthermore, an older discovery was found initially, before the foot bone. Mined from Croatia, it belongs to a 120,000 year old rib of a Neanderthal. Edward Odes quoted, "The modern external environment is doing things to our historical internal environment that we've never encountered before in our evolutionary history," The cases of colon cancer are ascending due to diets containing high saturated fat, and carcinogens in food preservatives were possibly the origin of a rise of stomach cancer in the late 19th century. Although we can carry cancer through our lineage and the genes they bestow, changes in our environment can call for an increase in a multitude of cancers (Strauss, 2021).

Hippocratic Physicians

The Hippocratic physicians came up with the earliest scientific theory about cancer (Hippocrate. Prorrhétique II, 1861). Following the 4th century BC, Hippocratic physicians began consultations regarding the treatment of cancer, proposing that the axilla, flank of the body, and thigh were the most laborious cancers to treat (Di Lonardo, Nasi, Pulciani, 2015). It was recommended that non ulcerated, or non fungated cancers should not be surgically treated. It was also believed that interference would mean death, and the patient would be fine not doing so. They also believed that the disease was associated with an excess of black bile (Di Lonardo, Nasi, Pulciani, 2015). The believed that the imbalance of the four body humours blood phlegm, yellow bile, and black bile resulted in cancer and other diseases. Additionally, it was thought that cancer developed whenever black bile became the dominant humour in the body (Di Lonardo, Nasi, Pulciani, 2015).

Archigenes of Apamea

Between the 1st and 2nd century AD, Roman medical author and practitioner Archigenes of Apamea invented a method to surgically remove cancerous growth (Hippocrate. Aphorismes VI, 1844). This was discovered from the Oribasius writings that were meticulously written. It was recorded that when the human body was diagnosed by some types of carcinoma they would have to be surgically removed because it would then be disconnected from the rest of the body.

Surgery could also be avoided with medications, as long as the cancer was diagnosed at an early stage. If the cancer is found at an advanced stage, it must be removed if the patient is strong enough to handle it (Daremberg Ch., 1862). It was additionally included that the circulation of blood needs to be impeded with a ligature in order to excise the tumor, some cases requiring all vessels surrounding the tumor to be binded. The position the patient should be in is the one most unchallenging to perform the surgery, and make an inscension. The Oribasius writings state that any nerves in the perimeter should be removed, and any membranes that have been attached to the growth needs to be scrapped off succeeding the precision. In the circumstance of a hemorrhage, the surgeon should avoid contact with nerves while sterilizing the region and placing a piece of cloth surrounding the bleeding part. Afterwards, a suture is to be performed. Ingredients such as bread, salt, and leek should be applied to the wound. The final instruction is to sanitize the area for the next 2-3 days, providing products that would assist with the healing process.

The Black Bile Theory

Physicians in Ancient Greece had a more limited knowledge regarding the human body - Hippocrates theorized that the human body was arranged of only blood, phlegm, yellow bile, and black bile. This black bile was supposedly the cause of cancer when it existed in a surplus amount in any section of the body. For the next 1,400 years, it was commonly believed to be true. Meanwhile in Ancient Egypt, cancer was believed to be produced by the Gods.

The Lymph Theory

Stahl and Hoffman introduced this theory that was made up of fermenting and degenerating lympth. This stirred up attention and support, backed up by physicians including 1700s Scottish surgeon John Hunter.

The Blastema Theory

Johannes Muller concluded that cancer was made up of cells rather than lymph in 1838, suggesting that cancer cells formed from blastema between normal tissues (Walter and Scott, 2016). It was his student, Rudolph Virchow, who discovered that all cells were derived from other cells (Walter and Scott, 2016).

How the Cause of Cancer was Discovered

Research concerning cancer is ongoing to this day, and we are able to cure some types. Finding a secure method of prevention or a cure is done with clinical trials and research studies. In the 20th century, the progress of cancer research was as efficient as ever American Cancer Society, 2014). Carcinogens, radiation therapy, chemotherapy, and more coherent methods of diagnosis were uncovered. Prior to these discoveries, Hippocrates black bile theory was replaced in the 17th century with the discovery of the lymphatic system, providing more understanding. In the late 19th century, Rudolf Virchow, German physician, changed the game with his research and recognition that all cells, including cancer, derived from other cells. More theories including ones surrounding trauma, and parasites sprouted. It was even thought that cancer spread "like a liquid". The discovery that the cause of stomach cancer being a worm was then awarded wrongfully with a Nobel prize in 1926. Karl Thiersch, a German surgeon, deduced the reason for cancer, concluding that it spread through malignant cells.

Chimney-Sweepers' cancer

Percivall Pott published the first report of cancer caused by occupational exposure in Chirurgical Observations Relative to the Cataract, the Polyplus of the Nose, the Cancer of the Scrotum, the Different Kinds of Ruptures, and the Mortification of the Toes and Feet (1775). He first perceived that men who worked as chimney sweeps in London had a high occurrence of producing skin sore with coal soot on their scrotums (Britannica, 2021). He then concluded that men who regularly exposed themselves to soot were at a higher risk to get scrotal cancer. This was an environmental factor that Pott reported was a cancer-causing agent. Pott's discovery opened conversations to prevent work-disease, coined as the chimney-sweepers' cancer (Britannica, 2021).

The First Recognized Resection of a Primary Brain Tumor

Rickman J. Godlee, a neurologist at the Hospital for Epilepsy and Paralysis, made this first acknowledgment of the primary brian tumor (Kirkpatrick, 1984). Godlee performed the first operation for a primary cerebral tumor who diligently recorded his medical performance, leading to further discussions and surgery on the cerebrum, and the development of modern neurosurgery. Unfortunately, Mr. Godlee's patient passed away after 28 days of meningitis and secondary complications, however a postmortem examination confirmed that no excised glioma remained (Kirkpatrick, 1984).

The First Recognition of Lung Cancer

1761 was when lung cancer was first identified. German physician Fritz Lickint prophesied a connection between smoking and lung cancer. The led to an anti-tobacco and smoking especially in Nazi Germany. Clinical studies linking enhanced smoking and lung cancer ensued, these statistics providing evidence to Lickint's discovery (Newswire, 2013). Richard Doll, Austin Bradford, and Richard Peto ran several tests, publishing the first study towards the idea. In 1964, the Surgeon general of the United States acknowledged that smoking was the prime cause of lung cancer. Who the exact person that discovered lung cancer itself is debatable, but either way, this enlightened people globally. (Newswire, 2013)

The First Recognition of Melanoma

Melanoma derives from the word 'melanose', an ancient Greek word for black, reflective, or the dark pigmentation characteristic of the condition (WHO, 2015). French physician Rene Theophile Hyacinthe Laënnec, the inventor of the stethoscope, discovered melanoma in 1816 (Roguin, Theophile, Laënnec, 2016).

Melanoma is only one type of cancer out of the myriad of cancers that exist among organisms. The initial discoveries of all the cancers we know to date have opened up many discussions throughout history. With no known cure, enlightenment is what will help us understand this disease and prevent, treat, and diagnose people. We can then progress down the road to decreasing the distressing effects cancer has on humankind.

Diagnosing Melanoma

Maria Ashraf

The diagnosis of melanoma comprises the process employed to reliably detect the presence of carcinogenic melanocytes and cancerous tumours (Lee & Kwon, 2019). It also consists of labelling and categorizing specific types of melanoma and appropriately giving patients access to required care (Lee & Kwon, 2019). Furthermore, it is invariably a complex and multifaceted process. Precisely, it is dependent on factors such as accurate melanoma categorization and utilizing various identification methods. Additionally, some melanomas may not have well defined symptoms making it challenging to detect and diagnose them (Pluta, 2011). Thus the considerable majority of the diagnostic process relies on frequent screening and accurate identification of melanoma. This chapter will describe the specifics of melanoma diagnosis and will also highlight secondary factors such as its significance and advancements.

To begin, the first keystone is understanding the diagnosis of melanoma relies on understanding the unique nature of melanoma. As defined in previous chapters, melanoma is adequately characterized as the cancer of the skin which primarily affects pigment producing melanocytes. Therefore, its accurate diagnosis is based upon the precise identification of melanocyte abnormalities on the surface of the skin. These abnormalities usually manifest in the form of moles or pigmented patches. As simple as this might sound, the unique nature of melanoma presenting on the skin as moles or melanin spots can be delusive. This tendency can render melanoma indistinguishable from normal skin conditions and be easily passed as moles or skin pigmentation. To tackle this issue, regular screening exams are used to detect any skin concerns or changes. These exams can be distinguished into two main types: self-screening exams and physician skin examinations. These screening systems are based upon several different factors related to the presentation of moles or pigmented areas of the skin. A widely used guideline is termed the "A-B-C-D-E's of Melanoma", which comprises of questions related to the asymmetry, border, colour, diameter, and evolution of moles or pigmented areas of the skin (CDC, 2021). While self-screening exams are beneficial in early diagnosis, they have limited efficacy due to issues with punctual compliance from patients. On the other hand, physician skin examinations are also shown to be effective in early diagnosis and a greatly reduced prognosis of diagnosed melanoma (Swetter et al., 2012 ; Tripp et al., 2017). The lack of using such screening mechanisms can lead to delay in detecting and properly diagnosing melanoma patients, which can lead to further complications and reduced treatment efficacy (Naik, 2021). Therefore, the need for early detection and diagnosis is imperative.

Furthermore, according to epidemiology studies done on the prevalence of melanoma, there has been an overall increase in the incidence of melanoma worldwide (Ferlay et al., 2021 ; Leiter et al., 2020). This rising prevalence is even more concerning considering the death rate of melanoma patients. While overall cancer death rates are stabilizing overtime, melanoma death rates are still continuing to grow at a gradual pace (Matthews et al., 2017). The prognosis and advancement of melanoma, like other forms of cancer, is highly correlated to the time period between detection and treatment. The sooner the cancerous tissue is discovered, the better the outcomes of treatment. For instance, it was found that early detection has been linked to better prognosis and reduced mortality. It was able to achieve 5-year survival rates of around 97% for early stages of melanoma (Holmes et al., 2018). Therefore, early detection and diagnosis are crucial in characterizing and directing the prognosis.

Categorizing the timeline and progression of melanoma is also a part of the diagnostic process, and is known as staging. Staging involves a description of the initial location and onset of melanoma in the body. It is utilized mainly to further categorize the melanoma and predict outcomes of treatment. A widely used system of staging in melanoma is known as the TNM system (Canadian Cancer Society, n.d.). It categorizes melanoma into 5 distinct stages, where each stage comprises factors such as specific cancer location, tumour thickness, and prognosis. The stages begin with stage 0, which is the precancerous stage. This is the stage that distinguishes the diagnosis of in situ melanoma from invasive melanoma. In the in situ stage or stage 0, cancer is only present on the superficial layer of the epidermis and is yet to progress through it (Bauman et al., 2018). This stage is a relatively low risk stage and mortality rates are extremely low with proper treatment (Bauman et al., 2018). On the contrary, the invasive stage of melanoma begins with stage I and II. These stages consist of varying levels of tumour thickness (ranging from 0.4mm to 4mm) and levels of ulcerations. The progression of these stages can lead into the locoregional stage -stage III- that constitutes the initial movement of the cancer to nearby lymph nodes and regions. This local growth is then followed by the precarious metastatic stage IV, which comprises the movement of the cancer to novel locations in the body (Canadian Cancer Society, n.d.). Overall, the main significance of staging in the diagnostic process is to characterize the melanoma and its progression which are crucial considerations in its treatment.

One of the major features of melanoma diagnosis is the use of histopathological categorizations of melanoma. There are four major forms of melanoma as defined by the Candian Dermatology Association. Firstly, one of the most prevalent types is Superficial Spreading Melanoma (SSM). It comprises almost 70% of all melanoma cases. This form of melanoma is categorized by pigmentation which can include shades of brown, black, and pink (Rastrelli et al., 2014). Additionally, the skin surface and other characteristics such as irregular boundaries and skin projections can distinguish it from other forms. It is also mostly diagnosed on the backs and the rear surface of the leg in males and females respectively (Rastrelli et al., 2014). An additional form of melanoma that is moderately prevalent around 4% - 15%, is called Lentigo Maligna Melanoma (LMM). In terms of pigmentation, this type presents as shades of black and brown patches. It can also be identified through irregular mole boundaries (Canadian Dermatology Association, 2021). This type is mostly found in older populations. Another subtype of melanoma is the Nodular Melanoma (NMM). This type can be categorized by the pattern of its growth. This melanoma grows inwards towards the skin surface. It is majorly categorized by darker pigmentations of black and brown (Canadian Dermatology Association, 2021). The last common form of melanoma is Acral Lentiginous Melanoma (ALM), which is highly prevalent in Asian and hispanic populations. Furthermore, apart from these four main types there are also additional rare variants of melanoma such as Desmoplastic Melanoma (DM) (Rastrelli et al., 2014).

Subsequent to understanding the various forms of melanoma and their clinical and histological presentations, is exploring the process of diagnosis. As previously discussed, the main diagnosis of melanoma begins with the observation of the skin and identifying the possibility of melanoma. This is an elemental step in the basic diagnosis of melanoma. Only after the possibility of a melanoma is identified, additional and more specialized diagnostic techniques are employed for further verification. The main purpose of these techniques is to accurately categorize melanoma based upon established guidelines and specifications. It involves the use of histopathological guidelines and diagnostic techniques. Firstly, a large part of melanoma diagnosis relies on the physician's own evaluation and judgement (Rastrelli et al., 2014). Therefore, it is crucial to have access to a reference system for the diagnosis of melanoma that can serve as a tool for making accurate diagnostic decisions. As discussed above, several screening tools are employed to detect the melanoma. Clinicians may also use other tools such as the 7-point glasgow diagnostic aid which helps characterize skin conditions (Walter et al., 2013). Additional to these are published clinical guidelines and handbooks that physicians can use to diagnose melanoma, for instance the Journal Of The American Academy Of Dermatology (JAAD) regularly updates and publishes "Guidelines of care for the management of primary cutaneous melanoma" to guide healthcare providers in melanoma diagnosis and management (Swetter et al., 2018). Furthermore, the physician's judgement is further supported by diagnostic tests and techniques. The diagnostic methodologies for melanoma have various types and purposes, but the main objective is to identify and classify the presenting melanoma in patients. This is primarily done by doing a thorough examination of the surface of the skin. However, in recent years, there has been an introduction of novel techniques, diversifying the current diagnostic procedures with additions in areas of technology and molecular diagnosis.

One of the most common types of diagnostic procedures is dermoscopy. A dermoscopy in the sense of melanoma is the utilizing microscopic tools to identify the type of melanoma (Holmes et al., 2018). It involves a careful observation of features such as the presence of specific patterns, colours, streaks, vessels, and structures (Rastrelli et al., 2014). There have been several studies that highlight the use and efficacy of dermoscopy for melanomas rendering it a good source for diagnosis (Holmes et al., 2018). This form of diagnosis can be helpful in staging the melanoma as well. For instance, utilizing the feature of patterns, it is known that melanomas have distinct and abnormal pigment networks (Rastrelli et al., 2014). A pigment network is a pattern formed by melanocytes and is seen to be a prominent diagnostic marker for melanoma. The presence of a dense pigment network can highlight that the cancer has progressed to a later stage. Similarly, other additional factors such as pigmentation can also be used to predict the progression of melanoma (Rastrelli et al., 2014).

An additional diagnosis technique employed to verify and support an initial diagnosis is skin biopsy. In cases where after a physical exam there is suspicion of the presence of melanoma, a small sample of the surface of skin under observation is removed and sent for further testing. There are three major types of biopsies used. Firstly, a method known as punch biopsies can be used to isolate cancerous moles (Canadian Cancer Society, n.d.). It involves using a circular blade tool to remove the suspicious cancerous area in a circular form. It is mainly utilized for cases where the entire area needs to be isolated. An additional form of biopsy is known as excisional biopsy. This type of biopsy utilizes a scalpel to remove the entire area under consideration. Lastly, another form of skin biopsy is shave biopsy (Canadian Cancer Society, n.d.). This involves shaving off a portion of the skin that is suspected to be cancerous using a blade or scalpel. All of these biopsies are used in different capacities depending on the type of melanoma suspected.

In addition to skin biopsy, in cases where there is suspicion of the spread of melanoma to the lymph nodes, lymph node biopsy can be performed. This mainly consists of the removal of some lymph node tissue or lymph nodes. There are two main types of lymph node biopsy. Firstly, fine needle aspiration (FNA) is utilized for the removal of discrete quantities of lymph node fluid or cells for examination of cancer presence (Canadian Cancer Society, n.d.) . Secondly, sentinel lymph node biopsy (SLNB) is utilized to measure the spread of the cancer into nearby regions and lymph nodes. The sentinel lymph node is the first lymph node attached to the tumour. A biopsy is done on this specific lymph node to determine the possibility and extent of spread (Canadian Cancer Society, n.d.).

As discussed earlier, the main objective of melanoma diagnosis is to properly evaluate and define the prognosis of the melanoma to make it easier to set up therapeutic objectives. For example, superficial melanoma is significantly different from nodal melanoma and has its own course of progression and treatment. Therefore, it is necessary to not only use diagnosis as a way to label the condition of patients but to also use it to further investigate and classify each patient case. This use of diagnosis extends into using several different testing mechanisms each with specific purposes. For instance, a diagnostic process may require the patient to undergo additional tests including imaging tests such as an magnetic resonance imaging (MRI), ultrasound, X-rays, etc. to further understand the prognosis of the melanoma (American Cancer Society., 2019). For example, in cases where the cancer is metastatic, additional tests will be required to determine the nature and scope of its spread. In the case of melanoma, additional tests such as an X-ray computed tomography (CT) scan may be performed to trace the spread of the melanoma.

While histological patterns and physical exams are popular techniques for diagnosis, over the years there has been an increase in melanoma diagnosis that is characterized by genetic changes and mutations. For example, around 5% - 12% of melanomas can be distinguished as hereditary melanomas compared to non-hereditary melanomas (Davis et al., 2019 ; Rossi et al., 2019). This not only distinguishes their prognosis but also has implications from the clinical outcomes and objectives of the patient. In terms of tracking hereditary and gene changes, genetic testing is a prevalent way to categorize melanoma. Some benefits of having a genetic diagnosis include a more personalized therapeutic approach and risk prevention in families (Stanford Health Education, 2020). Currently, one of the most well known mutations for melanoma is CDKN2A. In terms of diagnosis, CDKN2A testing in familial history of melanoma is highly suggested for better insight into possible risks and prognosis (Davis et al., 2019 ; Rossi et al., 2019).

Keeping up with the same theme of molecular based diagnostics, recently there has been an increase in utilizing immunohistochemistry for diagnosis. This type of diagnosis is particularly helpful where determining a histopathological based diagnosis is difficult or uncertain. There are two specific types of biomarkers that can signify the presence of melanoma: 1) melanocytic markers and 2) proliferative markers. Melanocytic markers are primarily utilized to evaluate the presence of cancerous melanocytes. Melanocytic markers such as Melin-A, S-100 can be used to stain cancerous melanocytes and verify the presence of melanoma (Weinstein et al., 2014). Whereas proliferative markers can prove to be effective in mapping out the proliferation and growth of the melanoma. The presence and growth of proliferate markers such as mitotic rates, Ki-67, etc. is an indication of worsening prognosis (Weinstein et al., 2014). Therefore, biomarkers can play a crucial role in diagnosing and predicting prognosis of melanoma.

In addition to the diagnostic techniques mentioned above, technology has also introduced new forms of diagnosis that are built upon diagnostic frameworks (Fried et al., 2020). Notably, self diagnosis apps like MoleScope and SkinVision are being utilized in diagnosis (Davis et al., 2019). The main goals behind their utilization being speeding up the diagnosis process and reducing the barriers to access. Additionally, automation of screening tools such ABCDE are also under development and early in practice. Lastly, other unprecedented approaches such as the use of Artificial Intelligence is also being incorporated into the diagnostic process (Curiel-Lewandrowski et al., 2019). All of these approaches, while fairly new, are being formulated on the basis of previously existing diagnostic criteria and frameworks.

Overall, the diagnosis of Melanoma is a crucial yet complex process. It begins with a primary focus on a rapid and accurate identification of the presence of Melanoma. This can be done through physical exams and screening. After this screening process, diagnosis focuses mainly on verifying the validity of the diagnosed condition. This is carried out through histopathologies and biopsies. Furthermore, diagnosis can also extend into additional testing utilized to undercover specifications and classify melanoma. With such importance in the characterization and prognosis of melanoma, diagnosis can be argued to be one of the basic foundations of effective treatment. In essence, the main goal of melanoma diagnosis is the accurate classification of the patient's condition and is one of the primary stages of defining the prognosis of melanoma. Moving on with the theme of prognosis, the next chapter will take a closer look at the impact of melanoma on patients.

Impact of Melanoma on the Patients

Aaliyah Mulla

The impact of receiving a cancer diagnosis can be devastating, and the effects of melanoma on a patient are widespread. Alongside physical symptoms, many patients experience significant emotional, social, and financial effects, all of which interact in a complex manner. The experience of melanoma varies from patient to patient and depends on a number of factors, such as a patient's previous health, financial situation, support system, disposition, the length and severity of the melanoma, and the quality of care available to a patient depending on where they live. This chapter will provide a general overview of what many patients reportedly experience.

In its early stages, melanoma does not typically cause serious physical symptoms. Patients may notice discolorations or growths on the skin that can cause itchiness, irritation, or soreness. The identification of such surface marks is typically the first sign that something is wrong, and usually leads to a person getting tested for melanoma or other skin cancers.

When caught early, melanoma can be removed with surgery to prevent it from spreading (American Cancer Society, 2019). This may seem like it should cause minimal disruptions to a patient's life. However, the stress and anxiety of receiving a cancer diagnosis can be significant (Cancer.Net, 2020). This is compounded by the worry that even after treatment, the cancer may return (Cheung et al., 2018). To account for the possibility of relapse, survivors of melanoma are monitored with regular followup appointments after completing treatment to ensure the cancer has not returned (American Cancer Society, 2020). For people who had early stage melanoma that was completely removed, these follow-ups could be every six to twelve months for multiple years, and typically involve a full physical exam (American Cancer Society, 2020). These follow-ups are in addition to self-exams conducted by patients, which involve searching for unusual lumps or discolorations on the skin (American Cancer Society, 2020). Moreover, survivors of melanoma may decide to make lifestyle changes to reduce the risk of melanoma returning, and to prevent the development of other cancers. This might include avoiding exposure to ultraviolet (UV) radiation (American Cancer Society, 2020), which can be done by avoiding exposure to the sun, particularly during midday (Government of British Columbia, 2016). Sun exposure can be limited by staying in the shade, wearing a wide-brimmed hat, and wearing sunglasses, among other precautions (Government of British Columbia, 2016). Other lifestyle changes might include eating a healthy and well balanced diet, exercising regularly, maintaining a healthy weight, and severely limiting or completely avoiding alcohol consumption (American Cancer Society, 2020).

While early stage melanoma can cause significant disruptions to daily life, the danger of melanoma spreading throughout the rest of the body is what makes it one of the most deadly skin cancers (American Cancer Society, 2019). This is why it is so important to treat melanoma early and continue monitoring for recurrence, even if early stage melanoma does not cause serious physical symptoms. The treatments used for more serious melanoma can cause a slew of physical symptoms, including extreme fatigue, nausea or vomiting, fever, chills, and rashes (American Cancer Society, 2019).

These symptoms can make it challenging, if not impossible, to complete what were once considered to be simple daily tasks. For instance, a parent may have trouble fulfilling regular caregiving duties (Cheung et al., 2018), such as assembling lunch for their child to take to school, driving their children to soccer practice, etc. Additionally, patients usually have to stop working, and some may find themselves without the energy or capacity to engage in activities or hobbies they find rewarding and enjoyable.

These lifestyle changes, combined with the stress of being diagnosed with melanoma, and the uncertainty surrounding recovery, can lead to a plethora of negative emotions, which can lead to poor mental health. Many patients report feeling some mixture of sad, anxious, depressed, scared, and angry (Cancer. Net, 2020). It can be scary to face the prospect of a shortened lifespan, and frustrating to be too tired to engage in social activities, or even regular daily tasks. In fact, debilitating fatigue is one of the most commonly reported symptoms of melanoma (Cheung et.al, 2018). This kind of fatigue cannot be solved with sleep (Melanoma Network of Canada, 2020).

However, the good news is that there are supports available for all of these symptoms, and treatments available for most. Research has shown that while sleep may not help, exercise tends to reduce cancer-related fatigue (Melanoma Network of Canada, 2020). Many patients find comfort in support groups, where they can discuss their feelings with others who are in similar situations (Melanoma Network of Canada, 2021). Support groups are spaces for people to connect with each other and realise they are not alone in the challenges they may be facing (Smart Patients, n.d.). Sometimes, they are run by oncology social workers (Cancer Care, 2021). These are professionals who are trained to connect cancer patients to resources that help them navigate the various facets of their illness (Akhilanda, n.d.). Oncology social workers are also equipped to help patients process their emotions, and they may suggest techniques to help cope with new lifestyles, such as checklists or new methods of prioritisation (Akhilanda, n.d.). The goals of support groups include offering hope to patients and encouraging feelings of empowerment (Cancer Care, 2021). Many groups include chats or forums for people to build community and discuss the highs and lows of their cancer journeys, outside of regularly scheduled group discussions (Cancer Care, 2021). Some organisations, such as the Melanoma Network of Canada, even offer support groups for caregivers (Melanoma Network of Canada, 2021). Support services of this kind are typically offered free of charge (Cancer Care, 2021; Melanoma Network of Canada, 2020). Similar to oncology social workers, there are also therapists who are specifically trained to work with cancer patients and can provide individual counseling (Good Therapy, 2018).

In contrast to the negative emotions that often accompany melanoma, many patients report positive feelings. This can include a feeling of calm, and of being more at peace with oneself (Cheung et al, 2018). One study that examined the effects of melanoma included interviews with patients. They reported a wide range of experiences, reflecting the diversity of the human experience (Cheung et al, 2018). One patient explained that through their cancer journey, they learnt to be kinder to themselves (Cheung et al., 2018). Another described the major swings between positive and negative emotions over the course of the disease, from the shock of receiving the initial diagnosis to the "elation" of discovering that a treatment had worked, to the disappointment of finding that the cancer had returned or spread (Cheung et al., 2018).

The physical and emotional effects of melanoma also have significant effects on a patient's relationships and social situations. As mentioned earlier in this chapter, the immense fatigue, as well as the other side effects associated with melanoma and its treatments, can hinder a person's ability to perform tasks they once performed with ease. This means they may not be able to take on the same share of household chores as they used to. They may also have a hard time engaging in hobbies and activities that they enjoy, such as painting or hiking. This inability to do daily tasks can change the dynamics of relationships. Patients often report feeling guilty about the burden their cancer has caused for their loved ones, especially those who are serving as caregivers (Cheung et al., 2018). Caregivers and loved ones may find themselves taking on a much greater share of household tasks, as well as helping to sort out the logistics of treatment and appointments, all while managing their own emotions surrounding their loved one's diagnosis. A melanoma diagnosis can be overwhelming for both the patients and their loved ones. However, it is important to know that a great deal of support services exist both for patients and their caregivers.

Sometimes, the stress of a melanoma diagnosis can put strains on relationships between spouses. Individual and couples therapy can help with this (Good Therapy, 2018). A lack of energy to engage in social events, such as weddings, parties, etc., can make it challenging to maintain social connections with friends and family (Cheung et al., 2018). Patients may feel guilty for being unable to support their loved ones. However, most loved ones are looking to do whatever they can to help patients, and many are very understanding about a patient's situation.

In the previously mentioned study, patients also reflected on these changes in social dynamics. Some reported that despite the immense challenges of melanoma, "there were definitely many, many blessings along the way," (Cheung et al., 2018). In some cases, this included significant support from communities, friends, and family (Cheung et al., 2018).

In contrast, some reported that their diagnosis was eye-opening for them in a different way. One patient reported that there were many people who they realised were not supportive of them, and that they broke off relationships with people they realised were not their true friends (Cheung et al., 2018).

In order to ensure patients are able to complete the tasks that are most important to them, they are often encouraged to make lists that prioritise what is most important (Akhilanda, n.d.). For instance, perhaps a patient hopes to be able to take their child to the park, do the dishes, as well as read a book they enjoy, all in one day. If spending time with their child is the most important of the three goals, they would do that first, and then read their book, and if they have energy left afterwards then perhaps they would attempt to do the dishes.

In general, cancer diagnoses tend to change people's outlooks on life, particularly when the cancer is advanced with a bleak prognosis. There are many stories of people redefining their priorities, reconnecting with loved ones, and reconsidering their values. For instance, the story of Ali Banat drew much attention when he was diagnosed with cancer in 2015 (MATW, 2020). Ali Banat was a self-made businessman who became a millionaire as a result of his successful company (MATW, 2020). After being diagnosed with cancer, he realised his time was limited and he wanted to reconnect with his faith (MATW, 2020). He donated his fortune to charity, and started a charity of his own so that he could leave behind a legacy that provided support to those in need (MATW, 2020).

In addition to the social, emotional, and physical repercussions of melanoma, many patients must also consider the financial costs associated with the condition. The financial burden of melanoma can be especially heavy for patients in the United States, where citizens must pay for their own healthcare, or rely on insurance coverage. The most obvious expense associated with melanoma is the cost of treatment. As with the other factors, the more advanced the melanoma, the bigger the costs. Melanoma that has been caught in an early stage can often be removed relatively easily in a single doctor's visit (American Academy of Dermatology Association, n.d.). More advanced melanoma, however, may require more extensive surgery, immunotherapy, targeted therapy, chemotherapy, or radiation therapy (American Cancer Society, 2019). Each of these have differing costs, and patients may also have to pay for drugs or treatment to help combat the side effects of the cancer treatments.

Aside from treatment costs, patients may incur other financial costs as a result of melanoma. For instance, if a patient must stop working, they risk losing their income. Depending on the severity of their melanoma, patients may have to apply for long term disability, go on unpaid leave, or leave the workforce on a permanent basis. There are services available to help patients navigate their new financial situations, and people, such as oncology social workers, who can help with paperwork and direct patients to the appropriate resources (American Academy of Dermatology Association, n.d.).

Moreover, patients may incur costs associated with lifestyle changes they may need to make in order to help manage symptoms or prevent cancer from recurring. For instance, a healthy diet is recommended for melanoma patients (American Cancer Society, 2020). According to one study, healthier diets are usually more expensive than less healthy alternatives (Rao et al., 2013). In this case, healthy diets were considered to be those high in vegetables, fruits, fish, and nuts (Rao et al., 2013). Unhealthy diets, on the other hand, were considered to be those high in processed foods, meats and refined grains (Rao et al., 2013).

Overall, a melanoma diagnosis can require a great deal of lifestyle changes — some major, and others relatively minor. Some of the other ways that melanoma may affect patients will be discussed below.

Depending on the type of treatment being administered, patients may find that food tastes different than normal (Ziembicki, 2021). Some patients report that food takes on a metallic taste, while others say food tastes salty or bitter (Ziembicki, 2021). Yet others say it tastes sweet or acidic, and some even say food has no taste (Ziembicki, 2021). Dieticians suggest ways to limit these effects. For instance, to combat a metallic taste, patients can add citrus juices to food and water (Ziembicki, 2021). Likewise, sugar combats saltiness or bitterness, and salt combats sweetness or acidity (Ziembicki, 2021). For foods that have lost their taste, dieticians suggest adding different herbs or sauces (Ziembicki, 2021).

During melanoma treatment, it is also important to stay hydrated, eat a balanced diet, and consume sufficient protein (Ziembicki, 2021). This can be tricky for patients who are struggling with nausea, loss of appetite, or a sore mouth (Ziembicki, 2021). For patients struggling with nausea or a loss of appetite, it is recommended that food be consumed at the time of day when they are most hungry, that patients eat small, frequent meals, and that they avoid eating and drinking at the same time (Ziembicki, 2021). For patients with a sore mouth or trouble swallowing, it is recommended to consume liquids, such as soups, and moist foods (Ziembicki, 2021).

Overall, the effect of melanoma on patients is as diverse as the patients themselves. While the illness no doubt brings immense challenges, many melanoma patients have been able to find hidden gems of positivity even amongst their pain. For every challenge faced by patients, there are people working on ways to make life easier and reduce the repercussions of the disease. As treatments and support services continue to be improved, there is great hope that melanoma will someday be easily treatable without causing major disruptions to daily functioning.

Melanoma Prevention

Jannat Irfan

As stated in the earlier chapters, melanoma is a tumour that is produced by the malignant transformation of melanocytes in the skin and bones (Heistein & Acharya, 2020). Melanocytes are derived from the neural crest called the melanomas and they usually occur on the skin and as said before in the other chapters, they can arise in other locations as well where the neural crest cells migrate such as the gastrointestinal tract and the brain (Heistein & Acharya, 2020). Melanoma is a type of skin cancer and once it develops in the melanocytes it can start to grow out of control and harm the health of a person (American Cancer Society, n.d.).

Melanoma is a less common type of skin cancer compared to other types of skin cancer. But unfortunately, it can be more dangerous than other types of skin cancer because it is much more likely to spread throughout the parts of the body if it is not brought to attention to the doctor or caught and if it is not treated in the early stages (American Cancer Society, n.d.). Melanoma is dangerous and can cause very serious consequences if one develops it. Therefore, the best way for people to not get melanoma is to protect themselves by prevention strategies. Prevention is of course better than treating melanoma since the person will not have any dangerous tumour, to begin with, if they take preventative measures. Prevention is a much better approach because while there are many discoveries that have been recently made for melanoma, a significant number of patients continue to be diagnosed with thick primary lesions and increased risk of metastasis and death (Curiel-Lewandrowski et al., 2012). In the USA, there is an increased incidence of melanoma in young women who are younger than 40 years of age (Curiel-Lewandrowski et al., 2012). However, the melanoma incidence has been largely stabilized in recent years especially in the age group 25-44 which is due to the effectiveness of primary prevention in younger age groups (Curiel-Lewandrowski et al., 2012). The older age groups are at higher risk of developing melanoma because of less knowledge about prevention and other health factors as well (Curiel-Lewandrowski et al., 2012).

One of the first steps that can be done in order to prevent melanoma is early screening and detection of possible melanoma since melanoma is one of the few types of cancer that has an increasing incidence rate (Rigel & Carucci, 2000). According to Curiel-Lewandrowski et al (2012), "Identification of the individuals that are at a risk of developing melanoma based on known predisposing risk factors and those at higher risk of dying from disease presents an opportunity to target the right individuals for more effective melanoma screening and the accessibility of the skin for both self and physician examination coupled with the ability to readily identify those at greater risk of developing and/or dying from a disease should support the concept of targeted melanoma screening." The survival rate of patients with melanoma is related to early detection since then it can be treated and prevented in the future. There are multiple factors that influence survival such as the thickness of the tumour which is the most important factor found out through various research studies (Rigel & Carucci, 2000). According to Rigel and Carucci (2000), "the five year survival rate for patients with early melanoma is defined as thinner than 1 mm and is 94% versus less than 50% for those with melanomas greater than 3 mm in thickness." But due to new treatment methods which are discussed in another chapter, early detection, and prevention strategies has seen an increase in the five year survival rate from approximately 40% in the 1940s to about 90% in 2000 (Rigel & Carucci, 2000). Prevention is known to halt melanoma according to several recent studies with an expected positive impact on mortality (Curiel-Lewandrowski et al., 2012). Nevertheless, there is still a lot of need to spread awareness about melanoma and teach the general public about melanoma, how and why is it caused, what treatments methods are used to treat it, and what preventative measures can be taken to ensure that nobody is at the risk of developing melanoma which has an increased mortality rate. There is still a significant need to take action on effectively implementing prevention strategies all around the world and especially in countries where there are no guidelines in accordance with recent studies (Rigel & Carucci, 2000).

As stated before, early detection, prevention strategies, and definitive therapy are necessary to minimize the risk of melanoma (Rigel & Carucci, 2000). It is possible and very easy to prevent melanoma through behavioural changes (Rigel & Carucci, 2000). Even with the new strategies that are currently being developed to treat advanced diseases such as melanoma, it is very important to put a strong emphasis on prevention in professional and public education (Rigel & Carucci, 2000). There are multiple proposed effective preventive strategies that should be implemented and strictly followed by the general public in order to prevent themselves from developing melanoma.

The first one is primary prevention which should be practiced by everyone. Primary prevention also means that the people do not have a developing melanoma condition but they are taking preventive measures and protecting themselves from the risk of developing a melanoma tumour. The best way to prevent yourself from getting melanoma is to reduce rates of sunburn by avoiding the sun during periods of peak UV radiation (Curiel-Lewandrowski et al., 2012). It is important to not be exposed to UV radiation for long periods of time since UV radiation exposure and the sun's rays play a big role in causing melanoma which is discussed in other chapters (Rigel & Carucci, 2000). Sunscreens are used as a protective health behaviour prevention strategy. It has been reported through various studies that the use of sunscreens reduces the risk of developing not only melanoma but other types of skin cancers as well (Rigel & Carucci, 2000). Sunscreens are very important because exposure to UV radiation is estimated to be one of the main causes of melanoma and the use of sunscreen daily blocks the UV radiation and prevents risks of developing melanoma (Sander et al., 2020). Therefore, the use of sunscreen regularly has been proven through various studies to reduce the incidence of both melanoma and nonmelanoma cancer (Sander et al., 2020). The Canadiana Dermatology Association as well as the American Academy of Dermatology recommends the general public to use sunscreen to prevent melanoma and other types of skin cancer (Sander et al., 2020). Sunscreens can protect the skin against UV radiation because sunscreens contain chemical and physical compounds which can be organic and inorganic respectively. These chemical and physical compounds work together to block the UV radiation which is the light with multiple wavelengths that are shorter than the visible light and shorter the wavelength the greater the potential for light radiation to cause biological damage (Sander et al., 2020). Sunscreen has filters in it that filter against the short wavelengths such as UVA1, UVA2, and UVB radiation and the chemical filters present in the sunscreen such as oxybenzone, avobenzone, octocrylene, and ecamsule are all aromatic compounds that are known for absorbing high-intensity UV radiation. This results in an excitation to the higher energy state and when these molecules return to their normal or ground state, the result according to Sander et al (2020). " is the conversion of the absorbed energy into lower energy wavelengths, such as the infrared radiation." Whereas physical filters in the sunscreen such as titanium dioxide and zinc oxide refract and reflect UV radiation away from the skin leading to skin avoiding contact with UV radiation and the sun's rays (Sander et al., 2020). Moreover, the zinc oxide also absorbs the UV light throughout most of the electromagnetic spectrum, preventing the development of melanoma since UV radiation is reflected back or absorbed by the chemical and physical filters (Sander et al., 2020). Evidence from multiple observational studies which assessed the effectiveness of sunscreen in preventing melanoma has shown that sunscreen does indeed prevent skin cancers including melanoma as well as photoaging (Sander et al., 2020). One of the high quality of evidence to support that sunscreen does indeed prevent melanoma is a randomized control trial of sunscreen for melanoma prevention which is called the Nambour trial (Wehner, 2018). The trial took place in 1992 and there were 1,621 randomly selected residents of Nambour, Australia that were selected to participate in the study aged between 25 to 75 years. All the participants were randomly assigned to daily or discretionary sunscreen application to the head and arms in combination with placebo supplements until 1996 (Green et al., 2011). The participants were observed until the year 2006 with various questionnaires, pathology laboratories and the cancer registry to ascertain primary melanoma occurrence (Green et al., 2011). The results of this randomized trial study showed that there were 11 new melanoma found in the daily sunscreen group

52

and 22 have been identified by the discretionary group which also demonstrated a reduction of the observed rate in those randomly assigned to the daily sunscreen use (Green et al., 2011). These results as well as the statistical evidence which the researchers found showed support and clear evidence that melanoma may be prevented and is preventable if sunscreen is used regularly (Green et al., 2011). Therefore, it is very important to wear sunscreen at all times regularly regardless of age, season, and location.

There are many other ways as well to prevent melanoma other than using sunscreen such as wearing protective clothing and shadow seeking behaviour (Wehner, 2018). Wearing protective clothing means covering up when going out in the sun and protecting your skin, for example, wearing clothes with long sleeves and pants (Protective Clothing, n.d.). Wearing protective clothes can shield people from the UV light and sun rays. Wearing protective and covering clothes is a very easy, accessible, and affordable way for people to shield their skin and protect it from the sun when sunscreen is not available (Protective Clothing, n.d.). The more the skin is covered, the better it is in terms of preventing melanoma development. Moreover, some clothing offers more protection from the sun such as some specific fabrics which are designed and tested for the purpose of measurable protection from the UV radiation and the sun's rays (Protective Clothing, n.d.). These fabrics are called the Ultraviolet protection factor and are used to measure the effectiveness of sun protective clothing The Ultraviolet protection factor also indicates what UV radiation penetrates the fabric and what rate of protection (UPF) it offers people (Protective Clothing, n.d.). For example, a shirt with a rate of 100 UPF allows only 1/100th of the sun's UV radiation to pass through the cloth to the skin and hence it keeps the skin very protected and reduces the risk of melanoma as well if worn regularly out in the sun (Protective Clothing, n.d.). Of course, the normal clothing also gives protection but UPF has been created scientifically and is scientifically engineered to protect skin from the UV rays and has been tested to ensure its efficacy (Protective Clothing, n.d.). However, the only limitation of UPF is that sometimes it can be too costly for people to afford (Protective Clothing, n.d.).

Another way to prevent melanoma is to avoid indoor tanning lamps and tanning beds (Le Clair & Cockburn, 2016). There is a misconception that indoor tanners which use artificial UV radiation are safe, however indoor tanning lamps and beds can be very dangerous for the skin and are not safe at all (Le Clair & Cockburn, 2016). According to Le Clair and Cockburn (2016), "The exposure to UV radiation from the indoor tanning machines has been shown to cause DNA damages in the skin cells and is associated with an increased risk of developing melanoma, and squamous and basal cell carcinomas and to add on, indoor tanning has also been associated accelerated skin ageing, ocular melanoma, immune suppression, as well as skin burns." As a result, the general public should be informed and be aware of the damages that indoor tanning devices can cause them since they are one of the leading causes of melanoma according to the World health organization (find the article - 10). In addition, research done by the World health organization (2017) shows that people who have used a sunbed at least once at any stage in their life are at a 20% higher risk of developing melanoma compared to people who have never used a sunbed (World Health Organization, 2017). Therefore, it is best to avoid tanning devices in order to prevent melanoma.

Another preventative measure that should be implemented in addition to the ones mentioned above is to have a healthy diet (Ombra et al., 2019). Nutrition plays an important role in all kinds of cancer and according to the World Cancer Research Fund, about 30-40% of all cancers can be prevented if there is a proper and a healthy diet intake, physical activity, and maintenance of correct body weight (Ombra et al., 2019). Moreover, various research studies have found that regular consumption of vegetables and fruits is directly associated with a reduced risk of melanoma (Ombra et al., 2019). It is important to eat fruits and vegetables to prevent melanoma because they are bioactive foods that contain compounds that have tumour suppressing properties (Ombra et al., 2019). Furthermore, a healthy diet means a healthy body, therefore a strong immune system to prevent the development of melanoma.

It is crucial to become familiar with one's own body and skin and be aware of all the changes that are occurring in the body and discussing those changes with a family doctor (Mayo Foundation for Medical Education and Research, 2020). It is important to notice and examine one's skin often for new skin growths or changes in freckles, bumps, moles, and birthmarks that were already present (Mayo Foundation for Medical Education and Research, 2020). Using a mirror, the ears, scalp, chest, trunk, arms, legs, genital area, feet, and face should also be examined and checked carefully in order to notice changes (Mayo Foundation for Medical Education and Research, 2020). Being updated regarding what is on one's skin is important because the early detection of melanoma is possible and there is a much higher chance of recovering if melanoma is detected early.

Finally, it may be concluded that melanoma is a very dangerous type of cancer which can cause death and should be prevented at all cost. The various ways in which melanoma can be prevented effectively are to avoid working directly in the sun especially during midday, to use sunscreen regularly, wearing clothes that do not expose your skin to the sun and the UV radiation, avoid using indoor tanning beds and lamps, eating a diet rich in fruits and vegetable and controlling body weight, and lastly knowing everything about oneself's body and the changes that occur on the skin. Melanoma is the cause of suffering and death of many people according to the World Health Organization (2017) and along with the patient who has melanoma, their family also sufferers because it is a dangerous life threatening condition. Hence, it is very important to take preventative measures to ensure a healthy life free of all types of cancer including melanoma.

Pathophysiology of Melanoma

Hafsa Binte Younus

As discussed previously, melanoma is the cancer of the skin. It is a form of tumor/cancer which is caused by the uncontrollable and harmful proliferation of melanocytes, which are a type of cell that is crucial for protection and pigment (Heistein & Acharya, 2020). Melanoma is one of the most common cancers in young adults aged 15-49 (Melanoma Network of Canada, 2021). In 2020, approximately 8,000 (4,400 males and 3,600 females) Canadians were diagnosed with melanoma skin cancer and it was estimated that around 1,300 Canadians would die from melanoma (870 males and 450 females) (Melanoma Network of Canada, 2021). Additionally, it is the 7th most commonly diagnosed cancer in Canada and has the highest mortality rate of all dermatological cancers (Melanoma Network of Canada, 2021). However, if detected in the early stages because it is often clearly visible on the skin, the survival rate for melanoma is high (Melanoma Network of Canada, 2021). This chapter of the book will explore the pathophysiology of melanoma.

Pathophysiology

Anatomical Pathology

There are 4 major types of melanoma that are classified according to the growth pattern. The first type of melanoma is known as the Superficial spreading melanoma. This type of melanoma constitutes approximately 70% of all melanoma diagnoses (Heistein & Acharya, 2020). This type of melanoma is usually flat with no elevations in the initial stages but may become irregular and elevated in later stages (Heistein & Acharya, 2020). The lesions for the superficial spreading melanoma on average are about 2 cm in diameter, with variegated colors, along with peripheral notches, indentations, or even both (Heistein & Acharya, 2020). This type of melanoma is mostly found on the back in males and legs in females (Emanuel & Cheng, 2013).

The second type of melanoma is known as Nodular melanoma (Heistein & Acharya, 2020). This type of melanoma constitutes approximately 15% to 30% of all melanoma diagnoses (Heistein & Acharya, 2020). The tumors of this type of melanoma are typically blue-black, however, in some cases, the tumor might have no pigment (Heistein & Acharya, 2020). This form of melanoma comes up as a rapidly enlarging nodule in which the tumor cells grow outwards and cause an upward epidermal invasion instead of horizontal epidermal invasion (Emanuel & Cheng, 2013). The tumor cells of this form of melanoma are round and epithelioid in morphology with hyperchromatic nuclei (Emanuel & Cheng, 2013).

The third type of melanoma is known as the Lentigo maligna melanoma (Heistein & Acharya, 2020). This type of melanoma constitutes approximately 4% to 10% of all the melanoma diagnosis and occurs on chronic sun exposed skin such as the skin of the scalp, face or neck (Heistein & Acharya, 2020). The tumors of this type of melanoma are often larger than 3 cm, are flat, and tan in color (Heistein & Acharya, 2020). They also have marked notching of the borders (Heistein & Acharya, 2020). These types of tumors begin as small freckle-like lesions but gradually increase in size as the melanoma progresses (Heistein & Acharya, 2020).

Finally, the fourth type of melanoma is known as the Acral lentiginous melanoma (Heistein & Acharya, 2020). This type of melanoma constitutes approximately only 2% to 8% of melanomas in Whites and 35% to 60% in dark-skinned people (Heistein & Acharya, 2020). This type of melanoma may appear on the palms and soles (Heistein & Acharya, 2020). They are often flat, have tan or brown stains, and have irregular borders (Heistein & Acharya, 2020). They are often found on the digits (including under nails), on the palms, and the plantar part of the feet. Subungual lesions (lesions under the nails) can be brown or black, with ulcerations in later stages (Emanuel & Cheng, 2013).

Development of Melanoma

To begin with, the development of melanoma may be related to various factors. Firstly, research has shown that family history is positively associated with the melanoma in about 5-10% of the patients (Heistein & Acharya, 2020). One is at about 2.2-fold higher risk of developing melanoma if at least one of their relatives has been affected with melanoma (Heistein & Acharya, 2020). Secondly, personal characteristics such as having blue eyes, having light skin and hair, having high freckles density, having multiple benign and/or dysplastic melanocytic nevi (moles), and being immune suppressant are all positively correlated with melanoma (Heistein & Acharya, 2020). Moreover, one of the leading causes of melanoma is being over exposed to ultraviolet (UV) radiation, whether it be from the sun or other artificial sources (Heistein & Acharya, 2020). UV radiation can cause the exposed human skin to tan or burn (Melanoma Network of Canada, 2021). Although both tan and burns are basically a result/ sign of damage to the underlying cells of the skin, a burn in particular is a marker of extensive damage that the normal DNA repair mechanism within a cell may not be able to repair (Melanoma Network of Canada, 2021). Sunburns are an indicator of the overexposure of skin to the ultraviolet radiation which increases the risk of skin cancer (Melanoma Network of Canada, 2021). Frequent childhood sunburns and exposure to toxic chemicals may also increase the risk of the development of melanoma (Swetter, 2019).

Melanoma can develop anywhere (Cole, 2020). It can develop in or near any precursor lesion (Cole, 2020). Some of the precursor lesions where the melanoma can develop include precursor melanocytic nevi such as the common acquired nevus, dysplastic nevus, congenital nevus, and cellular blue nevus (Heistein & Acharya, 2020). On the other hand, melanoma may also develop in healthy appearing cells too (Heistein & Acharya, 2020). Malignant melanoma which develops in healthy appearing cells with no prior evidence of precursor lesion is said to arise de novo (Heistein & Acharya, 2020). This type of melanoma accounts for more than 70% of the cases (Swetter, 2019).

How does Melanoma develop?

The process by which a normal melanocyte transforms melanoma cells is called melanogenesis (Swetter, 2019). However, the events related to this process and the sequence are not properly understood (Swetter, 2019). It is likely that the process of melanogenesis and the progressive genetic mutation of melanocyte involves multiple steps that first alter the cell proliferation, differentiation, and death abilities and second impact the susceptibility to the carcinogenic effects of ultraviolet radiation (Swetter, 2019).

More generally, the skin is made up of 3 main layers: Epidermis, Dermis, and Hypodermis. Between the dermis and the epidermis layers is a basement membrane that separates the 2 layers. Just above the basement membrane, there is a skin cell layer called the melanocytes. Melanocytes are important as they help protect humans from UV radiation and help with skin pigmentation. These melanocytes which are normally harmless, however after becoming mutated start to excessively divide and proliferate resulting in melanoma.

Melanomas have 2 main growth phases: The Radial Phase and The Vertical Phase. During the radial growth phase, malignant cells grow radially in the epidermis (Hasney et al., 2008). This growth face is circumferential in nature as the melanoma grows within the epidermis to the dermal-epidermal junction; it does not go beyond the junction (Hasney et al., 2008). However, at some point, a tumor in the radial phase undergoes a clonal change which grants them a survival advantage that allows them to enter the vertical growth phase as they pass through the dermal-epidermal junction (Hasney et al., 2008). In this phase, the melanoma grows into and invades the dermis and even grows beyond the dermis. In the vertical growth phase, the malignant/tumor also develops the ability to metastasize (Hasney et al., 2008).

Research also suggests that there are multiple pathways that result in melanoma pathogenesis (Swetter, 2019). A study done by Whiteman et al. compared the hypothesis that melanomas at different parts of the bodies have different associations with melanocytic nevi and solar keratoses (Whiteman et al., 2003). The researchers tested their hypothesis by performing a case-to-case comparative study of melanoma in Australian patients (Whiteman et al., 2003). They compared the melanocytic nevi count and solar keratoses between patients with superficial spreading (nodular) melanomas of the trunk, patients with superficial spreading/nodular melanomas on the head and neck, and patients with lentigo maligna melanoma (Whiteman et al., 2003). The results of the study showed that compared with patients with melanomas of the trunk, the patients that had head and neck melanomas were significantly less likely to have more than 60 nevi but significantly more likely to have more than 20 solar keratoses and a history of excised solar skin lesions (Whiteman et al., 2003). Moreover, patients with LMM were also less likely than patients with truncal melanomas to have more than 60 nevi and more solar keratoses (Whiteman et al., 2003). Through this data, it can be inferred that melanoma that developed in sun-protected skin areas, such as the trunk, was associated with high nevus (mole) counts and intermittent UV radiation (Whiteman et al., 2003). On the other hand, melanomas developed in areas where the skin was excessively exposed to the sun were associated with low nevus count and chronic exposure to UV radiation (Whiteman et al., 2003). Therefore, this research suggests that there are at least 2 pathways that can result in the development of melanoma: one associated with melanocyte proliferation and the other with chronic exposure to sunlight (Whiteman et al., 2003).

Biochemical and Genetic Pathology

DNA is the chemical in each of the human cells that make up their gene and genes are what control how the cells function (American Cancer Society, 2021). Cancers are often caused by DNA mutations that result in certain genes such as the oncogenes, genes that help cells stay alive, grow, and divide, to stay on and genes such as tumor suppressor genes, genes that keep the cell growth in check to ensure that the cell functions properly by repair DNA mistakes or by causing the cell to die at the right time, to turn off (American Cancer Society, 2021). These genes can be passed on from parents and inherited by the offspring. In some cases, cancerous melanoma genes can be inherited from the parents resulting in the child being at high risk of developing melanoma (American Cancer Society, 2021). The family inherited melanomas are often associated with the changes in tumor suppressor genes such as CDKN2A (also known as p16) or CDK (American Cancer Society, 2021).

However, mostly gene changes/mutations associated with melanoma are often acquired during a person's lifetime itself and are not inherited (American Cancer Society, 2021). In some cases, the mutation might be due to random error within the cell of the person and in other cases, the mutation may be caused due to external stimuli and exposures such as UV radiation (American Cancer Society, 2021).

Ultraviolet radiation is strongly associated with DNA mutations and is one of the major causes behind the development of melanoma (Puckett et al., 2020). One of the most common mutations in melanoma cells is the mutation of the BRAF gene which is an oncogene (Puckett et al., 2020). Other genes that may also be affected in melanoma cells include NF1, NRAS, KIT, CDKN2A, and TRET (Puckett et al., 2020).

The frequency of the mutations of BRAF and NRAS genes is highly associated with the pattern of sun exposures (Swetter, 2019). More BRAF mutations were more commonly found in intermittently UV-exposed skin, on the other hand, KIT mutations more prominent in chronically UV-exposed skin (Swetter, 2019). A meta-analysis by Lee et al shows that the prevalence of these mutations may also depend on the histologic subtype of the melanoma (Swetter, 2019).

Characteristics/Clinical Presentations

Finally, although the classical clinical representation of melanoma varies based on the type of melanoma, all the melanomas have characteristics that are assessed based on the ABCDE mnemonic, where each letter represents the following (Cole, 2020):

- **Asymmetry:** the two halves of the lesion are not identical but different from others (Cole, 2020).

- **Borders:** The lesion may have bleeding/undefined/ irregular borders which are not smooth and regular but uneven or notched (Cole, 2020).

- **Color:** The lession has several or changing colors (Cole, 2020).

- **Diameter/depth:** The lesion diameter keeps increasing or is greater than 6 mm (Cole, 2020).

- **Evolution:** The lesion changes in size, shape color, texture or even elevates (Cole, 2020).

Additionally, Melanomas may often also itch, bleed, ulcerate, or develop satellites (Cole, 2020).

Clinically, lesions are also classified according to their depth. The thickness and depth of the lesions are used as one of the important prognostic indicators for stage I and II tumors. The following depth ratios are used to classify the lesions:

- Thin - 1 mm or less (Heistein & Acharya, 2020)
- Moderate - 1 mm to 4 mm (Heistein & Acharya, 2020)
- Thick- greater than 4 mm (Heistein & Acharya, 2020)

Effect of Melanoma on the Body and Different Organs

Initially, the first signs and symptoms of melanoma include a change in an existing mole or the development of a new pigmented or unusual-looking growth on the patient's skin (Cole, 2020). These pigments would exhibit the characteristics associated with the acronym ABCDE explained above. Additionally, although most melanomas originate on the skin, they can develop on almost any organ of the body including the eye, brain, and lymph nodes (physiopedia, 2021).

Moreover, melanoma can be more serious as compared to other forms of skin cancers as they are metastatic (physiopedia, 2021). It has the ability to spread to the other parts of a patient's body even if it had originally originated on the skin (physiopedia, 2021). This usually happens in the advanced, metastatic, or stage IV of melanoma (physiopedia, 2021). The melanoma spreads when the tumor cells gain the ability to travel through the body's tissues, blood, or lymph nodes. They can implant themselves in any other part or organ of the body and keep growing. Therefore, they can move to any part of the body whether it be the lungs, liver, bones, lymph nodes, digestive system, and even the brain (physiopedia, 2021).

The effect of metastatic melanoma and how it makes a patient feel depends on where the melanoma spreads and the size of the tumor there. For example, if it spreads through the lymphatic system lymph nodes then the lymph nodes may get swollen and cause a lot of pain (physiopedia, 2021). The patient may also report other symptoms such as fever, night sweats, weight loss, or infection (physiopedia, 2021). On the other hand, melanoma may also spread through the circulatory system (bloodstream) to more distant sites such as the lungs, liver, other skin locations, or any other part of the body (physiopedia, 2021). If the metastasis moves to the lungs, then the patients might have symptoms such as coughing, shortness of breath, chest pain, or trouble breathing (physiopedia, 2021). If it goes to the liver, then the patient may experience indigestion, swollen belly, jaundice, or liver failure (Cole, 2020; physiopedia, 2021). On the other hand, metastases to the brain may result in the patient experiencing headaches, vomiting, unusual numbness in the arms and legs, or even seizures (Cole, 2020). Spread to the kidneys may cause pain and blood in the urine and spread to bones may cause bone pain or even result in the bones to break resulting in pathologic fractures (Cole, 2020). In rare cases, the release of melanin into the circulation could be so much that the patient may develop urine that is brown or black in color and have their skin turn into a diffuse slate-gray color (Cole, 2020). Moreover, if the melanoma has spread to the patient's remote skin sites, then the patient's skin may have the appearance of multiple blue-gray nodules (hard bumps). In general, a patient with melanoma often feels tired, has a reduced appetite and experiences weight loss (Cole, 2020).

Conclusion

Overall, there are many different types of melanoma, and although there are several factors that can result in the development of melanoma, most initial symptoms of melanoma involve the formation of abnormal pigments on the skin or the presence of one or more atypical nevus (mole). However, not all atypical moles or pigmentations are melanoma, therefore it is extremely important for individuals to get a comprehensive skin examination done by their dermatologist whenever they notice an abnormal pigment on their skin or if they have multiple or atypical nevi, a history of excessive sun exposure, or cutaneous skin cancer or melanoma. In the next chapter, treatments for melanoma and research associated with them are explored.

Research on Melanoma Treatment

Abeer Ansari

As discussed in earlier chapters, melanoma is a type of skin cancer with a high level of aggressiveness and severity and is therefore considered to be one of the most deadliest types of skin cancer. There are many different types of treatment options for melanoma and they have all evolved overtime since their efficiency has been decreased by the development of different resistance mechanisms that have forced researchers and clinicians to advance the treatment options. This chapter will start by providing a brief overview of the treatments, the importance of detecting the cancer during its early stages and what time of treatments and detection tools can be used, in-depth analysis and research on specific treatment options, side-effect of some treatment mechanisms and other research advancements that have taken place in the area of melanoma treatment (Domingues et al, 2018).

Overview of Treatments

There are various different treatments for melanoma and while some are generally more effective than others, the effectiveness and efficiency of the treatment depends on several factors such as the stage of the cancer, whether it is early and non-aggressive or much later in the stage where it is metastatic. Other factors include the risk of recurrence or cancer coming back, the location of the cancer, and how the treatment will impact one's psychological and emotional well-being (Canadian Cancer Society, n.d.). An important point to note is that oftentimes, patients may be offered more than one treatment option and would likely combine different treatments in order for the most effective recovery of the cancer.

Surgery is considered to be one of the main treatment options for melanoma and the type of surgery a patient's undergoes depends on the stage of the cancer and the risk of the cancer returning. Depending on these two factors, a patient may even be advised to undergo multiple types of surgeries. The different types of surgeries include: wide local excision, sentinel lymph node biopsy, complete lymph node dissection, reconstructive surgery and surgery for metastases (Canadian Cancer Society, n.d.). An in-depth analysis of these surgeries and the current research about them will be discussed later in this chapter.

Immunotherapy is another form of treatment that helps the patient's immune system fight off the cancer cells through the help of drugs. This type of treatment is usually used post-surgery in order to decrease the likelihood of the cancer coming back or for shrinking and controlling the growth of cancer cells in the case where surgery is not feasible (Canadian Cancer Society, n.d.).

Other therapies for melanoma include radiation therapy, chemotherapy and targeted therapy. Radiation therapy targets a small area of tissue using beams of radiation. This is often done either post-surgery to reduce the likelihood of the cancer returning back or is used as a palliative care treatment option for those who cannot undergo surgery but would like to limit the melanoma from advancing (Canadian Cancer Society, n.d.).

Chemotherapy is used to destroy the cancer cells by the use of cytotoxic drugs which are anti-cancer drugs. Patients undergo systemic chemotherapy in which the drugs are administered to the entire body and circulate throughout the body, attacking and destroying the cancer cells anywhere and everywhere in the patient's internal system (Canadian Cancer Society, n.d.). In contrast, regional chemotherapy is when the drugs are administered directly into a localized area such as the arm or the leg to target the cancer cells in that specific area of the body (Canadian Cancer Society, n.d.).

Lastly, targeted therapy is aimed towards special molecules or specific proteins that are found within or on top of the cancer cell. Such targeting using special drugs limits the abnormal growth of the cancer cell by altering its proteins and molecules (Canadian Cancer Society, n.d.).

Different stages of Melanoma and treatment

It is crucial to identify the development of this skin cancer during the early stages because once it reaches the metastatic state, the prognosis of treatment is extremely poor (Domingues et al, 2018).

In order to understand when different treatments should be used and the efficiency of each treatment, it is crucial to understand the different stages of melanoma cancer and how it progresses through each stage (Domingues et al, 2018).

A "stage" when referring to cancer is used to indicate how much the cancer has spread and where it started to spread from (Canadian Cancer Society, n.d.). This is the extent of cancer and is illustrated through the different stages. In melanoma cancer, there are 5 different stages ranging from 0 to 4. When describing these stages, clinicians usually use the 3 different descriptive words: early stage, locoregional and metastatic (Canadian Cancer Society, n.d.). Early stage signifies that the cancer has not spread yet and is only localized in the skin area where it initially began. Early stage includes the flow link stages : 0, 1A, 1B, 2A, 2B and 2C. Locoregional signifies that the cancer has spread but not too much. It has only spread to nearby lymph nodes and nearby skin areas or vessels. This includes stage 3 of melanoma. Finally, metastatic signifies that the cancer has spread significantly and is further away on the body from where it initially began. This includes stage 4 of melanoma (Canadian Cancer Society, n.d.).

In stage 0, the cancer cells are present only in the epidermis which is the outermost layer of the skin. During this stage, surgery is usually the best option to remove the cancer and the specific surgery type often used in this case is wide excision (Cancer.net, n.d.). In stage 1A, the tumour is not extremely thick since it has a thickness of only 0.8 mm and it does not result in any open wound or broken skin. However, in some cases, patients in this stage may have an open wound or broken skin but their tumour would not have more than 1 mm of thickness. In stage 1B, the tumour's thickness is between 1 mm and 2 mm without any ulcerations (Canadian Cancer Society, n.d.). For all sub-stages of stage 1, the cancer is usually treated with the surgical removal of the tumour and the nearby lymph nodes to prevent the likelihood of the cancer returning back or in case the cancer spreads to the nearby area but is not visible at the moment (Cancer.net, n.d.).

In stage 2A, if the tumour is less than 2 mm thick then it has ulcerations or it can be up to 4 mm thick without ulcerations. In stage 2B, when the tumour is up to 4 mm thick it is accompanied with ulceration which basically refers to broken skin or open wound while a tumour larger than 4 mm does not have ulcerations. In stage 2C, the tumour has exceeded the 4 mm thickness while creating broken skin and open wounds (Canadian Cancer Society, n.d.). For all sub-stages of stage 2, the cancer is removed surgically and nearby healthy tissue is also removed via surgery. Patients in this stage are also advised to undergo lymph nodes mapping in order to confirm the state of the early lymph nodes and the cancers of them getting the cancer. Some stage 2 patients also get treated with interferons which are signalling proteins, that reduce the chances of the cancer returning back (Cancer.net, n.d.).

In stage 3, the cancer starts to spread to nearby lymph nodes and the number of lymph nodes it affects, the amount of cancer present in the affected lymph nodes and whether or not the cancer has spread to vessels and other areas of the skin determines which substage of stage 3 the patient is assigned to (Canadian Cancer Society, n.d.). Since stage 3 melanoma spreads to different areas in the lymphatic system, it is important to locate all nearby lymph nodes and scan them for the presence of cancer (Cancer.net, n.d.). Based on the extent of the spread, if the cancer is able to undergo surgical removal, then surgery is the first option for treatment in this case. However, if surgery is not an option then advanced stage treatments need to be explored. It is recommended that patients in stage 3 who undergo surgery should also get post survey treatment by receiving immunotherapy or targeted therapy since it decreases the chances of the cancer returning back (Cancer.net, n.d.).

Finally, in stage 4, unfortunately, the cancer spreads to different parts of the body and damages other organs like the lungs and the liver. At this point, the cancer is called metastatic melanoma skin cancer (Canadian Cancer Society, n.d.). Individuals in stage 4 require advanced melanoma treatment as well as those in stage 3 whose cancer was not able to be removed surgically (Cancer.net, n.d.). During stage 4, the cancer spreads to other parts of the body that are farther away from the origin, including the liver, the lungs, bones, the brain, the gastrointestinal tract and other lymph nodes that are located at a significant distance (Cancer.net, n.d.). Treatment options usually include different types of therapy such as immunotherapy, targeted therapy or chemotherapy depending on the patient's personal preferences, how fast the cancer spreads, the location where the cancer has spread and the overall health condition and medical history of the patient (Cancer.net n.d.).

Brain Metastases

Brain metastases are when the cancer spreads to the brain. In the case of melanoma, the brain is the most common part of the body to which the cancer spreads and is associated with poor prognosis which means extremely low chances of recovery. Usually, patients whose cancer spreads to the brain live for about 6 months on average. The rate of recovery is extremely low since it is hard for drugs, such as those in chemotherapy, to reach the brain due to the blood-brain barrier which prevents foreign drug and medications from reaching the brain tissue. However, cancer cells also spread across brain tissue, making it an extremely risky case for the patient. Due to evolving research, there is still some hope for patients whose cancer has spread to their brain. These patients are recommended the following treatments: Radiation therapy, BRAF inhibitors and Immunotherapy (Cancer.net n.d.).

Radiation therapy is administered when there are only a small amount of tumours in the brain and it is administered through high-dose radiation stereotactic techniques since these techniques have been proven to be extremely effective in removing any tumours that are present in the brain. However, the limitation that comes with radiotherapy is the act that it is not efficient in preventing new tumours from developing. Another side-effect of radiation therapy for tumours in the brain is its negative impact on cognition and other brain processes due to the catastrophic radiation that is administered (Cancer.net n.d.).

Another treatment for brain metastases are BRAF inhibitors. Dabrafenib and Vemurafenib are drugs that have the ability to cross the blood brain barrier and enter the brain. This special ability makes these drugs a recommended treatment option for people with melanoma who have the BRAF mutation. Clinical trials have shown promising results since these drugs have resulted in a 40 to 50 percent shrinkage of melanoma turnouts that travelled to the brain (Cancer.net n.d.).

Lastly, immunotherapy is another effective treatment option for those with brain metastases due to melanoma. Clinical trials have studied the efficiency of ipilimumab, nivolumab and pembrolizumab and the most effective results were for the combination of nivolumab and ipilimumab therapy. However, since this treatment has considerable side effects, it may not be recommended for every patient (Cancer.net n.d.).

Interesting Research Findings on Melanoma Treatment

There are various different research findings in regards to treatment efficacy of melanoma and many if not all involved extensive clinical trials. A clinical trial by Doctor Georgia was performed to test the effect of combining two different therapies to treat patients who were previously untreated for advanced melanoma (Dabrafenib plus Trametinib for advanced melanoma, n.d.). Her trial included 423 patients who were diagnosed with advanced melanoma and had not received any treatment previously. The patients were randomly assigned to two different groups where one group received a combined therapy of dabrafenib and trametinib on a daily basis while the other group received only dabrafenib with a placebo. The results showed that the patients who received combined therapy had a 67 percent response rate and a 93 percent survival rate for 6 months while the patients who received dabrafenib and placebo had an overall 51 percent response rate and 85 survival rate for 6 months. Overall, the combination group had more promising results (Dabrafenib plus Trametinib for advanced melanoma, n.d.). Another clinical trial combined BRAF and MEK inhibitors in order to control the immune system's micro-environment to help fight the cancer cells in patients with melanoma and researchers are certain that the combination of dabrafenib and trametinib as well as the combination of dabrafenib and ipilimumab will bring promising results for combination therapies' efficacy (Smalley et al, 2016).

Throughout the 1990s and the early 21st century, there has been great progress and advancements in the field of research on melanoma treatment (Wróbel, 2019). In 1991, the Radiation Therapy Oncology group conducted a successful randomized clinical trial using radiation therapy on patients with melanoma and published their research. In 1993, it was discovered that the T-cells in the patient's immune system have the ability to recognize the melanoma antigens. In 1995, Bijay Mukherji and his team were able to successfully administer the first cell vaccination for melanoma. In 1996, high dose interferon for melanoma treatment was approved by the FDA after a successful clinical trial showed significant chances of both survival and low recurrence of the melanoma cancer in patients who were treated with high dose interferon. In 2002, the BRAF mutation in melanoma was discovered which brought a new perspective about melanoma etiology and has advanced treatment options in the present day for those who have the BARF mutation. In 2007, the first interferon trial came to an end and showed incredibly promising results which motivated the FDA to approve drugs such as ipilimumab and pegylated interferon, which as discussed earlier in this chapter, plays an important role in the treatment of melanoma. In 2013, after receiving promising results from programmed cell death and immunotherapy, the FDA approved dabrafenib and trametinib birth of which, as mentioned earlier, show promising results for combination therapy. Overall, there were a total of 198 registered clinical trials in 2015 and this shows an upward trend in the number of clinical trials that have been registered for melanoma each year over the last few years. These research advancements and ongoing clinical trials all hold a hopeful future for those suffering with melanoma (Wróbel, 2019).

Limitations on Melanoma Treatment

Jennifer Pham

Overview

Research into melanoma therapy has grown considerably in the last decade, with three different melanoma drugs approved by the Food and Drug Administration in 2011 (Lee et al., 2013). Currently, treatments for melanoma include surgery, radiation therapy, chemotherapy, immunotherapy and targeted therapy. However, despite the variety of available drugs, melanoma treatment is marked by numerous limitations. The challenges with treating melanoma can be summarized into two main areas: the efficiency of the drug, which includes the effectiveness of the drug and problems of drug resistance, and possible adverse events (AEs) (Domingues et al., 2018). This chapter will explore some current melanoma treatments and their associated limitations.

Surgery

Surgery is the main treatment for patients with stage 1-3B melanoma (Batus et al., 2013; Miller et al., 2016; van Zeijl et al., 2017; as cited by Domingues et al., 2018). While the procedures vary depending on the feature and location of the tumour (Canadian Cancer Society, n.d.), surgical treatments of melanoma mainly include wide excision, Mohs surgery, lymph node dissection and reconstructive surgery.

Wide local excision is the primary treatment for early stage, locoregional melanoma (Canadian Cancer Society, n.d.). It involves the removal of the tumour and some of the surrounding healthy tissue (Canadian Cancer Society, n.d.). Amputation may be needed depending on the location and the depth of invasion of the tumour (Canadian Cancer Society, n.d.). For instance, a tumour that has grown far into the skin and tissues on a finger or toe may require amputation of that finger or toe. Mohs micrographic surgery or MMS is a method that is commonly used in other forms of skin cancer but can also be performed on patients with melanoma (American Cancer Society, 2019). The procedure is a slow process of repeatedly removing thin layers of skin until a layer showing no sign of cancer is reached (American Cancer Society, 2019). Lymph node dissection is the removal of lymph nodes in the region around the tumour (American Cancer Society, 2019; Canadian Cancer Society, n.d.). Enlarged lymph nodes can be detected from a physical exam or imaging tests (American Cancer Society, 2019). If the lymph nodes are neither large nor hard, a sentinel lymph node biopsy can be done to detect whether lymph nodes contain cancer cells (American Cancer Society, 2019).

Side effects of surgery vary according to the type of surgical procedure, the location of the tumour and health conditions of the patient. Generally, some common side effects are pain, scarring, bruising, infection around the wound and numbness (Canadian Cancer Society, n.d.). However, lymphedema, a more serious side effect than the ones aforementioned, can develop after a lymph node dissection (American Cancer Society, 2019). Lymphedema describes the build-up of excess fluid that occurs when the removal of lymph nodes prevents proper drainage of lymph fluid. Lymphedema results in limb swelling that can lead to skin problems and increased risk of infection (American Cancer Society, 2019). Surgical treatments are not often recommended as a cure for melanoma that has metastasized (American Cancer Society, 2019). Once the cancer has spread to other organs, surgery is used to control it - such as by reducing the size or halting the growth of the cancer - rather than curing it (American Cancer Society, 2019).

Radiotherapy

Radiation therapy is not widely used in treating patients with melanoma (American Cancer Society, 2019). This method involves directing a beam of radiation to the area containing the tumour. It is mostly recommended to reduce the risk of cancer recurrence after surgery or as palliative therapy to relieve symptoms caused by the cancer (Canadian Cancer Society, n.d.-c). Radiotherapy can also be used in specific cases such as when tumour cannot be removed surgically or in the case of a rare type of melanoma called desmoplastic melanoma (Canadian Cancer Society, n.d.-c). Radiation can also be given for metastatic tumours in the skin, bone and brain (Canadian Cancer Society, n.d.-c).

Side effects of radiotherapy are mostly temporary and limited to the area being treated. They include skin redness and irritation, fatigue, hair loss in areas receiving radiation, sore mouth or throat for radiation in the head and neck area, and nausea for radiation at the abdomen (Canadian Cancer Society, n.d.-c). A more serious side effect of radiotherapy is lymphedema, which can occur if radiation is given in the groin or underarm area (Canadian Cancer Society, n.d.-c). In the case of treating metastasis to the brain, radiation can cause memory loss and headaches (American Cancer Society, 2019). Another limitation to radiotherapy is that the area surrounding the tumour is exposed to radiation, which can lead to damage in healthy tissues (Canadian Cancer Society, n.d.-c). Thus, doses of radiation have to be limited to prevent severe toxicity to the patients. In addition, as mentioned above, radiotherapy is not often used as curative care (compared to surgery) unless in certain cases. However, radiation is still valuable in palliative therapy and can be used in addition to some of the melanoma drugs named later in this chapter.

Chemotherapy

Chemotherapy was one of the first methods used to treat melanoma (Domingues et al., 2018). The cytotoxic drugs would attack and kill cells - such as by inhibiting cell division and growth processes (Canadian Cancer Society, n.d.-a). Chemotherapy can be subdivided into systemic chemotherapy, where drugs travel through the bloodstream to cells all over the body, and regional chemotherapy, where drugs are administered directly to the affected area (Canadian Cancer Society, n.d.-a). Systemic chemotherapy is often recommended if the melanoma has metastasized to other parts of the body, while regional chemotherapy may be used for recurrent melanoma that is limited to a limb (Canadian Cancer Society, n.d.-a).

Side effects for chemotherapy depend on whether systemic or regional chemotherapy was used. Common reactions to systemic chemotherapy include low blood count due to bone marrow suppression, hair loss, fatigue, sore mouth and throat, nausea, vomiting and diarrhea (Canadian Cancer Society, n.d.-a). Meanwhile, some side effects for regional chemotherapy are swelling, soreness and pain to the area where the catheter is inserted, hair loss and blood clots (Canadian Cancer Society, n.d.-a). While being one of the earliest treatments of melanoma, chemotherapy is not as commonly used today as immunotherapy and targeted therapy which are perceived to be more effective (Davis et al., 2019). Studies have shown that combinations of chemotherapy and other treatment methods have indicated good clinical responses but no improvement in the overall survival rate (Wilson & Schuchter, 2016; as cited by Domingues et al., 2018). Chemotherapy is also subjected to a resistance challenge in which cancer cells no longer undergo apoptosis despite being exposed to the drugs (Soengas & Lowe, 2003; as cited by Domingues et al., 2018). However, chemotherapy is still effective as palliative therapy for advanced melanoma.

Dacarbazine is the most commonly used drug in chemotherapy for melanoma. When used alone, dacarbazine shows poor activity (Blesa et al., 2011; as cited by Velho, 2012). For melanoma that has metastasized, the median survival after treatment with dacarbazine is 5-11 months and only 27% of patients survive after the one-year mark (DeVita Jr. & Chu, 2008; Lee et al., 2013; Rebecca et al., 2012; as cited by Davis et al., 2019). Despite these limitations, dacarbazine has remained the standard treatment in terms of chemotherapy for melanoma as other drugs have been even less effective and shown higher toxicity (Davis et al., 2019). In addition to monotherapy, dacarbazine is used along with other drugs. For example, combinations of bleomycin, vincristine, lomustine and dacarbazine (BOLD), and of cisplatin, vinblastine and dacarbazine (CVD) have been effective in treating metastatic melanoma. However, both BOLD and CVD did not improve patient survival (Serrone et al., 2000; as cited by Velho, 2012). Another combination of dacarbazine, carmustine, cisplatin and tamoxifen (DBDT) showed a better response in respect to dacarbazine alone; but DBDT presented increased toxicity and no improvement to the overall survival rate (Chiarion Sileni et al., 2001; as cited by Velho, 2012). Side effects of dacarbazine in particular include low blood platelet count (thrombocytopenia) and low white blood count (leukopenia) (Velho, 2012).

Temozolomide is another widely used chemotherapy drug to treat melanoma. Temozolomide is mostly helpful in cases of brain metastasis due its ability to cross the blood brain barrier (Chiarion Sileni et al., 2001; as cited by Velho, 2012). Thus, the decision between dacarbazine and temozolomide may depend on whether brain metastasis has occurred. In terms of effectiveness, according to a Phase III study performed by Middleton et al., the median survival time for advanced metastatic melanoma after treatment with temozolomide is 7.7 months (Middleton et al., 2000; as cited by Velho, 2012). Similar to dacarbazine, there are many clinical trials testing the effect of temozolomide when used in combination with other melanoma therapy drugs as well as radiotherapy. Some side effects specific to temozolomide are reduced levels of lymphocytes (lymphopenia), headaches, and constipation (Velho, 2012).

Immunotherapy

In recent years, immunotherapy has gathered a growing interest as an adjuvant treatment for metastatic melanoma. Immunotherapy involves administering medications to aid the immune system in fighting the cancer (Canadian Cancer Society, n.d.-b). This treatment is often used to reduce the risk of cancer recurrence or to control a tumour that cannot be resected (Canadian Cancer Society, n.d.-b). There are two types of drugs used in immunotherapy for advanced melanoma: cytokines and immune checkpoint inhibitors. Cytokines have a wide variety of antitumour mechanisms. For example, they can work by stimulating the cell of the immune system to recognize tumour cells (Canadian Cancer Society, n.d.-b). On the other hand, immune checkpoint inhibitors are monoclonal antibodies that work by binding to checkpoint proteins and blocking them to allow the immune system to attack the cancer cells (Canadian Cancer Society, n.d.-b).

Side effects of immunotherapy vary depending on the type of medications used. Cytokines can induce side effects such as fever, chills, aches, loss of appetite, fatigue, nausea, vomiting, diarrhea, rashes and low blood counts (Canadian Cancer Society, n.d.-b). The side effects of immune checkpoint inhibitors are similar and they include headaches, fatigue, rashes and diarrhea (Canadian Cancer Society, n.d.-b). Immune checkpoint inhibitors can also cause problems related to the liver, such as yellow skin and eyes, and problems related to the thyroid and lungs, such as coughing and difficulty breathing. (Canadian Cancer Society, n.d.-b) While immunotherapy is considered to be more effective as a treatment for advanced melanoma than the traditional therapies mentioned above, challenges with immunotherapy also include cancer recurrence and varying success rate (Domingues et al., 2018). The cancer cells may have primary or acquired resistance to immunotherapy drugs that could suppress an immune response, such as by evading recognition by T-cells (Domingues et al., 2018). Tumour heterogeneity is another contributing factor to resistance against anticancer drugs (Domingues et al., 2018). Thus, immunotherapy has been used with other treatment strategies such as chemotherapy and targeted therapy.

One of the main immunotherapy cytokines used to treat melanoma is interferon alfa-2b (IFN α-2b). The drug is mainly used for early stage melanoma or advanced melanoma that is limited to a specific region to reduce the risk of cancer recurrence (Canadian Cancer Society, n.d.-b). As an adjuvant treatment after therapy, the 5-year survival rate is 44% and the proportion of patients who did not experience cancer recurrence past the 5-year mark is 32% (Hancock et al., 2004; cited by Velho, 2012). IFN-α is also used in combination with other drugs. For example, a regimen of IFN-α and dacarbazine can suppress the growth of a tumour by regulating pericytes, which affect tumour vasculature (Liu et al., 2011; cited by Velho, 2012). However, when administered at high doses, IFN-α can lead to adverse effects such as fever, fatigue, chills and autoimmune problems (Velho, 2012). High doses of IFN-α-2b is also linked to depression (Canadian Cancer Society, n.d.-b).

These effects can be offset through using peginterferon α-2b (Peg-IFN), which is a pegylated form of IFN-α-2b (Velho, 2012). Peg-IFN has a higher efficacy than IFN-α-2b because it is able to stay in the bloodstream for a longer amount of time (Harris & Chess, 2003; as cited by Domingues et al., 2018). However, it also has some side effects which include low hepatotoxicity, low number of neutrophils (neutropenia), anemia and skin rash (Zarour et al., 2015; as cited by Domingues et al., 2018). More serious common side effects observed are lymphopenia and low concentration of sodium in the blood (hyponatremia) (Zarour et al., 2015; as cited by Domingues et al., 2018).

Another cytokine used in melanoma immunotherapy is interleukin-2 (IL-2). This drug can be used to control metastatic melanoma, and treat metastatic melanoma as well as tumours that cannot be operated on (Canadian Cancer Society, n.d.-b). IL-2 works by increasing T-cell division and function (Velho, 2012). As treatment for metastatic melanoma, IL-2 has shown a 6% complete response rate and a 10% partial response rate (Atkins et al., 1999, 2000; as cited by Velho, 2012). While high doses can increase efficacy, they also increase toxicity (Velho, 2012). Adverse effects of IL-2 treatments can range from low blood pressure (hypotension), swelling of limbs due to fluid buildup, rapid heart rate (tachycardia), irregular heart beat (cardiac arrhythmia) to reversible multisystem organ failure (Bhatia et al., 2009; as cited by Velho, 2012).

For immune checkpoint inhibitors, a commonly used drug is ipilimumab, which targets the protein CTL-associated antigen 4 (Canadian Cancer Society, n.d.-b). The drug is used to manage metastatic melanoma and tumours that cannot be removed with surgery (Canadian Cancer Society, n.d.-b). According to a

Phase III study by Hodi et al., ipilimumab alone has an overall survival of 10.1 months. Another study found that the 3-year survival rate is 20% and patients did not experience recurrence of the cancer (Hodi et al., 2010; as cited by Velho, 2012). However, the clinical response rate for the treatment is only around 10% (Merlino et al., 2016). Similar to many of the drugs mentioned in this chapter, there has been research into using ipilimumab in combination with other anticancer drugs such as Peg-IFN, IL-2 and nivolumab. Another challenge with ipilimumab is its adverse effects, which include dermatological reactions (such as erythematous, reticulated or maculopapular rashes), diarrhea and other gastrointestinal effects (such as colitis), thyrotoxicosis, hepatitis, pancreatitis and neuropathies (González & Ratner, 2016).

Nivolumab is also a widely used immune checkpoint inhibitor in treating melanoma. Unlike ipilimumab, nivolumab targets the PD-1 proteins (Canadian Cancer Society, n.d.-b). It can be used to completely remove stage 3 or 4 melanoma and reduce the chance of cancer recurrence (Canadian Cancer Society, n.d.-b). The drug can also be used to control metastatic melanoma and tumours that cannot be resected (Canadian Cancer Society, n.d.-b). A study indicated that 73% of patients survive past 12 months after being treated with nivolumab (George et al., 2017; as cited by Davis et al., 2019). Akin to other drugs mentioned, nivolumab can be used in combination with other treatments. For example, a regimen of nivolumab and ipilimumab have shown a progression-free survival of months, meaning the cancer did not worsen for 11.5 months after the regimen was administered (Byrne & Fisher, 2017; Palmer et al., 2011; as cited by Davis et al., 2019). The side effects for a nivolumab treatment are the same as those of ipilimumab and present a challenge for patients.

Targeted Therapy

Research into the formation and development of melanoma has paved the way for the recent developments in targeted therapy. This approach aims to limit the growth of cancer cells and the damage to healthy cells (Canadian Cancer Society, n.d.-d). Two common the current targeted therapies for melanoma are v-Raf murine sarcoma viral oncogene homologue B1 (BRAF) inhibition and mitogen-activated protein kinase (MEK) inhibition. Approximately 70% of cutaneous melanoma is due to mutations in the BRAF gene which are thought to stimulate melanoma cell growth and proliferation (Domingues et al., 2018). Thus, BRAF inhibition drugs target these mutated proteins to disrupt pathways that lead to uncontrolled cell growth and division (Canadian Cancer Society, n.d.-d). MEK inhibition has the same purpose as proteins expressed from the BRAF gene work by activating MEK proteins (Canadian Cancer Society, n.d.-d).

A key advantage of targeted therapy is that it is unlikely to harm healthy cells (Canadian Cancer Society, n.d.-d). This approach also has fewer side effects compared to some of the other treatment methods such as radiotherapy and chemotherapy (Canadian Cancer Society, n.d.-d). The side effects of targeted therapy vary depending on the type of drugs used but can include fatigue, nausea, vomiting, fever, skin problems, eye problems, muscle pain, swelling, diarrhea and abnormal liver function (Canadian Cancer Society, n.d.-d). While targeted therapy offers promising results, drug resistance can develop as well. Secondary resistance has been observed in a large number of patients within a short period of time after treatment (Chapman et al., 2011; Hodi et al., 2010; Rebecca et al., 2012; as cited by Davis et al., 2019). New mutations, changes to tumour microenvironment and epigenetic factors can arise and contribute to drug resistance for targeted therapy (Merlino et al., 2016). However, there is ongoing research to look into other proteins that can be used for targeted therapy drug development.

Vemurafenib and dabrafenib are two of the drugs used for BRAF inhibition targeted therapy. In a Phase III study by Chapman et al., the 6-month survival rate after treatment with vemurafenib was 84%, compared to the 64% rate after treatment with dacarbazine (Chapman et al., 2011; as cited by Velho, 2012). Dabrafenib has also shown a higher progression-free survival rate compared to dacarbazine (Hauschild et al., 2012; as cited by Velho, 2012). Furthermore, both drugs work for melanoma metastasis in the brain (Velho, 2012). Similar to some of the drugs before, higher doses of vemurafenib and dabrafenib correlate to higher toxicity and can lead to adverse effects such as rashes, light sensitivity, diarrhea, nausea, fatigue, joint pain (arthralgia) and alopecia

(Chapman et al., 2011; Soffietti et al., 2012; as cited by Velho, 2012). There has been drug resistance observed in treatments with vemurafenib and dabrafenib.

On the other hand, some common MEK inhibition drugs include trametinib and cobimetinib. Trametinib has shown better progression-free survival and overall survival rates when used as a monotherapy compared to chemotherapy (Flaherty et al., 2012; as cited by Velho, 2012). However, MEK inhibitors can also be used in combination with BRAF inhibitors. For example, a combination of cobimetinib and vemurafenib indicated good results by improving the progression-free survival rate (Hoeflich et al., 2012; Niezgoda et al., 2015; as cited by Domingues et al., 2018). Some common side effects for regimen using both MEK and BRAF inhibitors include fatigue, chills, fever, nausea, vomiting and diarrhea (Livingstone et al., 2014; as cited by Domingues et al., 2018;). Resistance to MEK inhibition also occurs. For example, tumour heterogeneity has been observed as one of the reasons that the cancer continues to progress after treatment with BRAF and MEK inhibition (Merlino et al., 2016).

Although drug resistance has presented many challenges in treating melanoma, these challenges open opportunities for research. Aside from BRAF and MEK inhibition, there has been ongoing research in other types of inhibitors such as vascular endothelial growth factor (VEGF) inhibitors, mTOR pathway inhibitors and cyclin-dependent kinase (CDK) inhibitors (Domingues et al., 2018).

Conclusion

This chapter discusses some of the common methods of treating melanoma and their associated limitations. Surgery seems to be the main treatment for melanoma, and other treatments are used in combination with surgery, or in cases of advanced melanoma and when surgery is not an option. Development of melanoma treatment seems to hang on the balance between efficacy of the method and its adverse effects. While some drugs show promising results, they can lead to serious side effects. Another key challenge with melanoma treatments is the potential of drug resistance which has been observed in some of the more effective treatments such as immunotherapy and targeted therapy. Combinations of drugs and treatment types have been used to overcome problems with drug resistance and to increase efficacy. These limitations also opened many avenues for future research into the development of melanoma treatments.

Epidemiology of Melanoma

Angelina Lam

Melanoma, also known as malignant or cutaneous melanoma, is a skin cancer in which melanocyte cells proliferate uncontrollably (American Cancer Society, 2019a). These melanocytes are responsible for producing melanin, which is the pigment that is responsible for varying skin colours. When cancerous, they usually create black- or brown-coloured tumours, which may spread to other parts of the body in advanced cases (American Cancer Society, 2019a). Although melanoma is rare, since it comprises 1% of all skin cancers, melanoma is still dangerous because it causes a leading portion of skin cancer deaths (American Cancer Society, 2021). In 2021 alone, 7,180 Americans are predicted to die due to melanoma (American Cancer Society, 2021). Despite melanoma occurrences being more prevalent as people age, it is also one of the most common cancers to develop in adults aged 20-39 (American Cancer Society, 2019a, 2019b). With these statistics, it is important to further research and understand melanoma in order to reduce incidences and deaths in the population. Thus, the epidemiology of melanoma is one avenue of investigation used to reduce the harm caused by this disease.

An Introduction to Cancer Epidemiology

The study of epidemiology has long been important in the field of health sciences. Epidemiological research focuses on disease occurrences in different populations and groups; this information is then used by professionals to guide actions for prevention and management (Coggon et al., 2003). The epidemiology of cancer takes this concept and utilizes it in the context of cancer diagnoses and prevalences. This field analyzes information through a variety of viewpoints, ranging from descriptive epidemiology that focuses on the time, whereabouts, and characteristics of people that acquire the disease, to analytical epidemiology that looks into specific groups for comparisons (Centers for Disease Control and Prevention, 2012; Oliveria et al., 1997). It also includes clinical epidemiology, which examines the implementation of screening and prevention programs and their impact on the population (Nazario et al., 1995). By evaluating disease trends, the importance of certain risk factors, and the effect of public health programs, epidemiology proves to be an important aspect in cancer research (Nazario et al., 1995).

Using these methods, cancer epidemiology aims to understand how various factors such as age, sex, and other lifestyle choices may play a role in increasing or decreasing the risk of cancer (Phillips, 1969). Through research, factors such as tobacco chemicals, alcohol consumption, diet, and physical activity have been found to be correlated with cancer risk; other factors such as occupation-related carcinogens, microorganisms or viruses, and hormonal factors can affect cancer risk as well (Oliveria et al., 1997; Peto, 2001). Furthermore, genetics can play an important role in cancer cases as certain risk factors may be passed on within a family (Peto, 2001). Genetic studies have found that certain alleles or mutations may heighten the risk of acquiring cancer, especially if they are certain changes in oncogenes or tumour-suppressor genes (Peto, 2001). This demonstrates the importance of genetic epidemiology in cancer research. Additionally, one must not forget that social determinants play a role in health and cancer outcomes (Toporcov & Filho, 2018). The epidemiology of cancer helps to bring all these factors together for review to help determine the disease's risks and causes.

As cancer epidemiology continues to develop, it will help reveal and substantiate the factors that may contribute to causing cancer, which will help inform preventative initiatives (Oliveria et al., 1997). Its discoveries will help illustrate disease distribution, and consequently, its control plans (Toporcov & Filho, 2018). Thus, it is important to have cancer surveillance programs in order to acquire data for analysis. This data collection can extend into treatment and palliative care as well, which helps to refine treatment strategies. The epidemiological process is multifaceted and has many considerations; however, in-depth assessments will help bring clinical applications to aid disease prognoses

(Toporcov & Filho, 2018). Epidemiological research helps quantify the likelihood of disease, highlights at-risk populations, and assists health professionals in understanding the effectiveness of treatments (Nazario et al., 1995). Overall, these positive impacts illustrate the importance of epidemiology in cancer research.

The Epidemiology of Melanoma

The importance of cancer epidemiology cannot be overlooked; thus, with a disease such as melanoma, it is beneficial to analyze it from an epidemiological point of view. As previously introduced, melanoma is one of the deadliest skin cancers (Matthews et al., 2017). It can also manifest itself in the eye and on mucosal surfaces; however, 90% of melanoma cases are on the skin and are the focus of many epidemiological studies, including this chapter (Horrell et al., 2015). The exact type of skin melanoma can be broken down into different categories, with some common ones including: superficial spreading melanoma (SSM), nodular melanoma (NMM), lentigo maligna melanoma (LMM), and acral lentigo melanoma (ALM) (Rastrelli et al., 2014). SSM occurs the most often as it comprises 70% of melanoma cases, while LMM, ALM, and NMM comprise smaller percentages that fluctuate depending on the population's ethnicity, age, etc (Rastrelli et al., 2014). This is estimated to be 4% to 10%, 15% to 30%, and <5% for LMM, NMM, and ALM, respectively (Carr et al., 2020). However, together, the total incidences of melanoma worldwide are still concerning.

Each year, over 100,000 people are affected with skin melanoma worldwide, while the age-standardized rate of incidence was reported to be 2.8-3.1 per 100,000 people (Ali et al., 2013; Horrell et al., 2015). Although the global prevalence of some cancers are decreasing, melanoma incidences are still increasing as, in the world, it is found to be the 19th most frequently-occurring cancer (Ali et al., 2013; Rastrelli et al., 2014). Within the United States, melanoma was the fifth most prevalent cancer for men and the sixth most prevalent for women as of 2014; its incidence has been reported to be 1 in 63 for Americans and 1 in 50 for those residing in the Western world (Rastrelli et al., 2014). It tends to grant high five-year survival rates (98%) when detected and removed at early stages of the disease, which become poor once the cancer metastasizes in the body (Carr et al., 2020; Horrell et al., 2015). By 2020, melanoma incidences were still increasing, with a 270% increase spanning the past three decades and a 44% increase in the past decade (Carr et al., 2020; Skin Cancer Foundation, 2021). In the United States alone, 100,350 cases of melanoma are predicted to be diagnosed, with 6850 deaths occurring in conjunction with incidences of the disease (Siegel et al., 2020). Despite that, some studies have reported a decreasing incidence of melanoma in recently-studied cohorts, accompanied by a large reduction of mortality from the disease. Mortality rates have dropped around 5% to 7% per year depending on the age group, where previously, mortality rates were declining around the 1% to 3% mark or in some cases, even increasing. This decrease is attributed to the approval of new drugs; however, epidemiological study likely also played a role in analyzing the incidence, risk factors, and treatment of the disease (Siegel et al., 2020).

The epidemiological study of melanoma encompasses many different factors; one example is the age of the individuals who are diagnosed with melanoma. Throughout the entire world, the age-standardized incidences of melanoma increase with age, with peak melanoma incidences occurring in those aged above 75, or those aged between the seventh decade and the ninth (Matthews et al., 2017). The increase in incidences starts at those aged 25 to 30 and above, with a majority of the cases occurring in the senior population of those aged 65 and older (Ali et al., 2013; Rastrelli et al., 2014). At age 25, the rate of incidence is also found to be linearly correlated with age, where it then plateaus and slows around age 50 (Rastrelli et al., 2014). The median age of diagnoses is found to be at age 57, while the median age of melanoma-related death is found to be 68 from a four-year study (Rastrelli et al., 2014). These statistics are helpful because they will allow for the examination of those at-risk for melanoma based on age-related factors.

In addition to age, sex is also a factor in the prevalence of melanoma. Generally, men are diagnosed with melanoma more often than women; this is shown with an age-standardized rate across all men of 29.2 in 100,000 in the United States, in comparison to the parallel figure of 17.3 in 100,000 for women (Matthews et al., 2017). Other studies have quoted that the ratio of melanoma in women to men is 2:3 (Rastrelli et al., 2014). Generally, however, sex is a factor that is highly interdependent on other factors such as age and geography. In terms of age, at younger ages, women are more likely to develop melanoma in comparison to men (Matthews et al., 2017). However, this figure contrasts to the incidences of melanoma above the age of 40, in which men are more likely to develop the disease than women (Matthews et al., 2017). In fact, at ages above 75, the ratio of melanoma in women to men is almost 1:3, being cited as 47.3 in 100,000 and 145.6 in 100,000 annually for women and men, respectively (Rastrelli et al., 2014). Mortality rates were also found to increase for men while it was decreasing for women; however on the other hand, melanoma was the second most frequently occurring cancer and highest cause of cancer-related death for women in their 20s (Horrell et al., 2015). In terms of geography, it was also found that in places with high incidences of melanoma, there was not much difference in the frequency of melanoma between the sexes, while the differences were more pronounced in regions of low incidence (Marks, 2000). Thus, this data shows that sex is an important factor in melanoma occurrence.

To further delve into geography impacting melanoma incidence, this factor deals with the varying risk of ultraviolet (UV) radiation between different geographic locations (Matthews et al., 2017). Previous studies of melanoma mortality rates have shown an increasing figure for populations closer to the equator and vice versa, as there are varying amounts of UV radiation as latitude changes. This was coined a latitude gradient and is generally seen worldwide as well as within countries, since countries closer to the equator present higher age-standardized rates of melanoma (Matthews et al., 2017). Certain countries also have higher recorded rates of melanoma. Australia was reported to have incidence rates of 37 in 100,000 in 1998, which was the top of the list for incidences; in 2012, of 184 recorded countries, New Zealand and Australia ranked highest for melanoma cases with 35.8 and 34.9 in 100,000 cases annually, respectively (Chang et al., 1998; Matthews et al., 2017). Despite numbers fluctuating by study and by year, the Australian and New Zealand regions still are reported to have one of the highest incidences of melanoma as it is one of their top reasons for cancer-related mortality (Carr et al., 2020; Horrell et al., 2015; Rastrelli et al., 2014). Europe follows behind Oceania, with countries such as Switzerland, Netherlands, Denmark, Norway, Sweden, and the United Kingdom, amongst others, reporting age-standardized incidences of 14.6 to 20.3 in 100,000, annually (Matthews et al., 2017). However, there are drastic differences in the European region as countries such as Greece report incidences as low as 2.2 in 100,000, annually (Ali et al., 2013). North America is also one of the regions with higher incidence rates, citing 13.8 cases per 100,000 people, annually (Matthews et al., 2017). Not only do all the listed regions have high incidences; what they also have in common is that many countries report gradients that show more cases in higher-latitude regions of low-latitude countries (Matthews et al., 2017). Examples of this North-to-South differential are seen in Scandanavian countries such as Denmark and Switzerland (Matthews et al., 2017). Other research has also found gradients from East to West, and this may be attributed to differences in lifestyle or data collection (Ali et al., 2013). Through solid epidemiological data collection, researchers can better understand melanoma risks in all countries around the world.

In addition to geography, varying ethnicities between countries and regions also play a role in the epidemiology of melanoma. As melanoma is a cancer of the skin, it largely affects those without skin pigmentation, in the form of melanin, to protect the skin (Matthews et al., 2017). Thus, Caucasian populations with a fair skin tone are highly affected as they do not have a melanin layer to shield them from UV radiation. Depending on if it is UVA or UVB radiation, the melanin of darker skin tones helps to reduce 23% to 50% of the UV radiation responsible for cancer (Matthews et al., 2017). This is a reason why the incidence for melanoma is 2.6%, 0.58%, and 0.1% for Caucasian, Hispanic, and African American populations, respectively (Carr et al., 2020). The Surveillance, Epidemiology, and End Results (SEER) program from the United States also reported similar trends, with decreasing incidence rates in the following order: Caucasian individuals, Indigineous peoples, Hispanic individuals, Asian individuals or Pacific Islanders, and African American individuals (Matthews et al., 2017). Additionally, individuals with a fair skin tone also may be affected at higher rates as pigmented moles, or melanocytic nevi, is a risk factor for melanoma (Marks,

2000). It is, however, important to note that all of these factors still interact with others such as age, sex, and geography (Matthews et al., 2017). Additionally, that information only covers incidence rates; appearances of the disease are also complex, and the disease presents dissimilarly in different ethnicities (Horrell et al., 2015). This is shown as Caucasian individuals often are diagnosed with SSM, while individuals with darker skin tones are often diagnosed with LMM and are diagnosed later into the disease, resulting in a higher mortality rate. This is also shown as African American and Asian individuals, for example, develop melanoma on regions that are not commonly exposed to the sun, while Caucasian individuals develop melanoma on regions of the body that are (Horrell et al., 2015). Overall, this understanding from epidemiology again helps to identify populations that are at risk of disease to help better prevent, detect, and treat melanoma.

To further explore the presentation of the disease, epidemiology also takes a look at the anatomical regions where the disease manifests. This factor is highly reliant on previously mentioned topics such as sex, geography, and ethnicity. As for sex, melanoma can be found on areas of low sun exposure for both, but regions far from the equator such as Australia show high incidences of melanoma in areas of high sun exposure (Matthews et al., 2017). Different regions present different statistics, as Australia and North America have seen increased melanoma on the trunk of women, while Europe saw an increase in melanoma on the lower limbs of women instead (Marks, 2000).

Another study also reported women more commonly finding melanoma on the upper and lower limbs, while men find it on the back (Carr et al., 2020). Generally, head and neck melanoma rates have remained constant for select countries, with a rise in cases on the trunk (Marks, 2000). However, the shoulder, upper arm, and back still remain the most common areas of melanoma growth (Matthews et al., 2017). This information helps the understanding of melanoma development and highlights areas of concern for UV radiation education.

Another interesting aspect of research is molecular and genetic epidemiology among those affected with melanoma. Molecular epidemiology focuses on the interactions between genetic factors and the environment of individuals, examining genetic factors down to the molecular mechanisms to understand how diseases occur (Honardoost et al., 2018). Genetic epidemiology takes a broad view of how certain genetics of individuals and populations can interact with environmental factors to cause disease as well (Khoury, 1997). These two fields are important as they can attempt to trace melanoma incidences to specific molecular or genetic interactions. For example, some studies have investigated the cause of certain melanoma mutations and the link between different types of UV exposure, mutations, and cancer types (Bauer, 2010). Genetic research has tried to find correlations between certain genes, melanocyte senescence, and skin pigmentation (Bataille, 2003). Although this research may be complex, increasing advancements are helping researchers find correlations between molecular processes, genetics, and the environment to better discern high-risk individuals.

As discussed, age, sex, geographic location, ethnicity, and genetic factors all can affect whether an individual will develop melanoma. However, there are still more epidemiological risk factors; these include lifestyle UV radiation exposure, environmental exposures, health factors, medical factors, and more (Carr et al., 2020; Horrell et al., 2015). For example, it is known that UV radiation exposure has been linked to melanoma incidences, and UV radiation is a known carcinogen (Ali et al., 2013; Carr et al., 2020; National Institutes of Health, 2011; Sample & He, 2018). Sun exposure in early childhood or youth ages may increase the incidence of melanoma in later life, but it is also argued that total exposure is a comparatively more substantial factor (Ali et al., 2013; Horrell et al., 2015). However, multiple studies say that intermittent exposure results in a heightened risk for melanoma (Ali et al., 2013; Carr et al., 2020; Horrell et al., 2015; Marks, 2000; Rastrelli et al., 2014). In addition, sunburns have been linked to melanoma as a history of sunburns have been seen to increase the risk, especially if they were severe (Ali et al., 2013; Carr et al., 2020). Since UV exposure is a risk, this makes it so that tanning beds also pose a melanoma risk. This is shown in research, which states a 16% to 20% increased risk due to tanning activities (Carr et al., 2020). UV exposure from occupations such as welding are also of concern, as well as other occupational exposures to pesticides or heavy metals, since they are linked to melanoma (Carr et al., 2020;

Horrell et al., 2015). This information is particularly useful because it helps professionals identify UV-related risks and for the public to learn what to avoid.

As previously mentioned, other melanoma risks include health and medical factors. As touched upon earlier in this chapter, genetics factors can make an individual more susceptible to melanoma. This includes previous family history of the disease (Horrell et al., 2015). An individual previously being diagnosed with melanoma also heightens their risk of redeveloping melanoma, especially with continued exposure to UV radiation (Ali et al., 2013; Horrell et al., 2015). Certain genetic variations can also cause molecular mechanisms to not be as efficient at repairing UV-radiation damaged DNA, which can encourage melanoma incidence (Carr et al., 2020). This makes it evident that medical history plays a large role in melanoma susceptibility. For example, individuals with xeroderma pigmentosum, an inherited skin disorder, have defective DNA repair mechanisms that make them more susceptible to UV radiation damage (Horrell et al., 2015; MedlinePlus, 2020). Additionally, immune deficiency can increase the chance of melanoma by 50%, while UVA therapy or medication (Psoralen) for skin conditions such as psoriasis can increase sensitivity to UV radiation, heightening the chances of disease. The presence of nevi, or moles, can also result in cancer risk as melanoma can emerge from large or dysplastic nevi (Horrell et al., 2015). Through this, it is important to be attentive to certain populations who may have health or medical histories that make them more susceptible to disease.

Future Trends

Collected epidemiological data is of utmost importance as it allows for the prediction of future trends in melanoma incidence and mortality. Studies state that melanoma diagnoses have increased while mortality rates have not followed, which may suggest improved detection in the population or more cases of localized, non-malignant melanoma being detected (Matthews et al., 2017). However, higher case counts of melanoma with thicknesses and subtypes across the board can also be interpreted as increasing melanoma incidences worldwide, which is not a positive sign (Matthews et al., 2017). Examining the data of certain countries, Europe has not seen a plateau of melanoma cases and neither has the United States; however, Australia has begun to see a levelling of melanoma incidences in young adults (Garbe & Leiter, 2009). Mortality-wise, death rates seem to be decreasing, although the median age of death may increase due to longer life expectancies. Despite this, positive outcomes such as levelling melanoma rates in young adults and lowering mortality rates can be attributed to interventions such as public health campaigns on UV radiation, better screening techniques, etc (Garbe & Leiter, 2009). This provides insight on what strategies are working in the fight to lower melanoma incidences worldwide.

Therefore, epidemiological studies have made it possible to implement effective programs and public health interventions to reduce melanoma's impact on human health. This data helps researchers understand how external factors, intrinsic factors, surveillance, changing incidence rates, treatment, mortality, socioeconomic status, and other variables could affect melanoma rates (Koh & Geller, 2011). Future goals from analyses include maintaining up-to-date and accurate registry programs for melanoma statistics, facilitating public health campaigns that would reduce UV exposure or use, implementing effective screening programs based on those at risk, amongst others (Koh & Geller, 2011).

Conclusion

In conclusion, epidemiology is imperative to the future of melanoma research and control. Through epidemiological work, it is possible to improve education, create more efficient screening programs, and use collected data to better implement public health policies regarding UV exposure (Koh & Geller, 2011). Early detection of melanoma saves lives; prevention methods can be further developed to best serve the at-risk population and mitigate the effects of the disease (Koh & Geller, 2011; Matthews et al., 2017). Successful programs in countries such as Australia demonstrate the validity of epidemiological work to better focus on and target initiatives to those most likely to develop melanoma (Matthews et al., 2017). Overall, the cumulative discussions in this field goes to show that epidemiology is foundational to the future of melanoma research.

Effects of Melanoma on National, Global & Public Health

Paige Breedon

Melanoma is a severe form of skin cancer that is known to be deadly. Specifically, its destructive nature is due to its ability to spread to other organs if it is not treated at an early stage (Melanoma, 2020). Melanoma originates as a melanocyte involved in skin pigmentation or pigment-producing cells in the outer layer of skin. These cells then proliferate and form a tumour (Melanoma, 2020). Melanoma has many impacts on every patient that it inflicts and overall has many national, global, and public health-related repercussions.

Firstly within Canada, melanoma has had a significant impact. For instance, a study in Canada aiming to evaluate the burden of 18 skin and subcutaneous diseases, including melanoma, on public health from 1990 to 2017 found an increased prevalence and impact of melanoma. Specifically, the study used incidence, prevalence, mortality, years of life lost, years lived with disability, and disability-adjusted life years to quantify the various skin diseases' impact on public health (Bridgman et al., 2020). The results of this study pertaining to melanoma specifically found an increase in the rate of incidence and prevalence from 1990 to 2017 by 68% and 81%, respectively (Bridgman et al., 2020). This statistic emphasizes how melanoma is continuing to affect Canadian people. The rationale behind this increase could be due to factors such as the ageing population and climate change. However, these rationales will be discussed later on in great detail. Additionally, out of the 18 skin and subcutaneous diseases evaluated in the study, melanoma accounted for the most years of life lost by representing 60% of years of life lost from skin disease (Bridgman et al., 2020). This statistic shows just how immense the impacts of melanoma are even when compared to other skin diseases. The study also emphasized that although increasing trends in Canada seem significant, these trends are consistent with other countries' melanoma rates, such as the United States, Sweden, Australia, New Zealand, Norway, and the United Kingdom (Bridgman et al., 2020). Specifically, these countries have experienced age-standardized malignant melanoma incidence rates in light-skinned individuals, rising by approximately 3% each year from 1982 to 2011 (Bridgman et al., 2020). The comparison between Canada and other countries demonstrates that the burden of melanoma is more significant than Canada and has many global implications. However, based solely on the study, Canada has had an increase in the incidence and prevalence of melanoma. Melanoma, when compared to other skin ailments and diseases, is one of the most significant contributors to adverse life effects such as years of life lost. The increasing trends and significant repercussions associated with melanoma justify action and awareness within the nation of Canada to achieve a less disease-burdened future.

Moreover, the 2020 Melanoma Skin Cancer Report provided by the Global Coalition for Melanoma Patient Advocacy and Euro Melanoma clearly illustrates the significant global prevalence and relevancy of melanoma. Specifically, over the last decade, the annual cases of the deadliest form of skin cancer (melanoma) have increased by almost 50% to an overwhelming 287 000 cases and approximately 60,000 deaths per year ("2020 Melanoma," 2020). The latest data from WHO suggests that the number of cases will continue to increase by approximately 20% by 2025 and 74% by 2040, thus creating an even greater cause for concern ("2020 Melanoma," 2020). This report, like many statistics circulating about melanoma, clearly demonstrates an increased prevalence of melanoma and increasing trends that will continue to burden future generations. For this reason, it is crucial to consider the immense relevancy of melanoma and prioritize preventative efforts, awareness, and research into treatments in order to benefit the quality and longevity of life and to prevent the economy from the immense effects.

Based on the provided data, it is clear that melanoma is a national and global threat to public health. Specifically, the significant effects of melanoma across national, global, and public facets are consistently related to the quality of life, finances, and life span. Firstly, melanoma as a severe skin disease has many significant effects on an individual's day-to-day life and, therefore, affects those individuals' ability to operate and contribute to society. Thus the personal impact of melanoma escalates quickly to impact entire communities and countries. Also, the cost of melanoma is high and relevant from the individual through the national to the global level. These costs can be both direct and indirect and put a tremendous strain on health care systems throughout the world. Furthermore, the prevalence of melanoma is consistent with a shortened lifespan, therefore impacting the global population. Overall, melanoma can affect national, global and public health by hindering productivity and quality of life, economically straining health care systems, and controlling populations.

Also, when discussing melanomas' large-scale impact, it is essential to understand that this skin disease will only worsen due to the ageing global population, climate change, and ozone depletion. All of these factors ultimately make the population more vulnerable to melanoma occurrences. Firstly, considering that the global population is shifting to a more senior demographic, the incidence and prevalence of melanoma will only escalate because older people have a greater chance of having and experiencing adverse effects of melanoma (Rees et al., 2016). Also, considering environmental factors such as climate change and ozone depletion can increase the global population's U.V. radiation exposure, which is a key driving factor in causing melanoma where specifically 86% of melanoma is associated with U.V radiation exposure, the amplification of melanoma's impact will continue (Memon et al., 2021). Overall, the national, global and public health effects of melanoma will only increase due to an ageing population, climate change, and ozone depletion. Therefore time must be invested in combating these driving factors to shield future generations from the mighty burden of melanoma.

Firstly, melanoma can seriously impact one's quality of life and wellbeing. These impacts can also transgress to affecting loved ones, members of the community, and when evaluated from a distant perspective, the entire human population from one extent to another. Melanoma, like many other diseases, has many implications that impact an individual's day-to-day lifestyle. Specifically, melanoma requires time spent in a hospital setting being treated and monitored. It can also have adverse effects on one's physical health and, therefore, become a significant part of their lives, having the ability to affect their mental and emotional state. Overall the effects of the disease can be very traumatic and extensive for many patients and, when evaluated on a large scale, can impact significant groups of people. As occurrences of melanoma increase, the impact on the general public's quality of life from a vast perspective will continue to increase and serve as a prime example of the immense capability of this disease.

A quantitative, non-interventional study had patients

recruited from seven provinces across Canada to analyze the humanistic effects of melanoma on a patient. Individual interviews were conducted where patients answered questions and shared their experiences with advanced melanoma. The most common disease-related physical impact pertained to sleep problems, primarily due to anxiety and stress (Cheung et al., 2018). Also, for many patients, their condition was debilitating; specifically, a third of patients reported limb impairment due to tumours or neurological symptoms when their melanoma spread to their brain. Patients also experienced effects on their emotional wellbeing; for instance, around 50% reported their sense of self-change amidst their diagnosis and experience with melanoma (Cheung et al., 2018). Approximately 66% of patients felt worried about their uncertain future; concerns mainly included patients worrying about their future lifespan, quality of life, and the impact their illness and or potential death could have on their loved ones (Cheung et al., 2018). Also, 90% of patients declared that melanoma had an impact on their social relationships. Overall, the study demonstrates how real patients in Canada specifically feel about how melanoma has impacted their physical, emotional, and social wellbeing (Cheung et al., 2018). Ultimately, melanoma has a significant impact on the quality of life for the patient, the patient's family, the surrounding community, and as melanoma increases in prevalence, perhaps larger groups of the population.

Furthermore, the economic burden of melanoma is significant for everyone involved starting from the individual and proceeding to affect entire countries. For the individual, costs come at a direct and indirect level. Direct costs include those associated with hospitalization, prescriptions, and treatments. Indirectly, costs for that patient can include loss of productivity and income. These costs can significantly affect a patient and their family, and as melanoma continues to occur, more groups of people will be financially burdened.

A retrospective, longitudinal survey intending to analyze health care cost patterns relating to melanoma in the United Kingdom, France, and Italy utilized data from hospitalization costs, hospice care, and outpatient visits (Johnston et al., 2012). The results of the study found hospitalization costs highest in France (6262 euros per person compared to 3225 euros in the U.K. and 2486 euros in Italy) (Johnston et al., 2012). Regardless, all three countries have a significant price associated with hospitalization due to melanoma, thus clearly illustrating the immense economic repercussion associated with this disease. Also the results noted that the variation in these costs is due to varying intensity and frequency of patient utilization of these resources (Johnston et al., 2012). Overall, the study emphasizes the importance of prioritizing melanoma as a global issue due to its economic repercussions for the individual.

Similarly, due to melanoma's immense cost, certain socioeconomic groups are more susceptible to melanoma wrath based solely on financial security and capability ("The effect," 2018). For instance, a study intending to determine the increased odds of being diagnosed with stage four melanoma compared to stage zero or one due to race, insurance, and additional indicators of socioeconomic status identified 225 280 individuals from the National Collision Database ("The effect," 2018). The chi-square test determined that race and ethnicity, insurance type, income and education were strongly associated with the diagnosis of advanced-stage melanoma ("The effect," 2018). In conclusion, the study found that Hispanic white and African-American patients who did not have private insurance and had decreased levels of income and education had a strong correlation with an increased chance of being diagnosed with stage four melanoma ("The effect," 2018). Overall, the study demonstrates that due to the financial burden associated with melanoma, large demographics of people are more vulnerable to the adverse implications of this skin disease. These demographics exist at the national, global, and public health levels.

Moreover, although melanoma may not directly impact an individual financially, the national costs associated with the pressure that melanoma, especially considering its uprising trends, put on the healthcare system can be costly for the average taxpayer. Although the extent to which melanoma burdens an individual as a patient versus a taxpayer varies depending on the economic infrastructure of their country's health care system, there will always remain an economic impact of melanoma at the national, global and public level. Specifically, in the

United States, among studies examining various stages of melanoma, annual medicare costs, which is health care funded by the U.S. federal government (i.e. taxpayers), can range from $44.9 million to $932.5 million (Guy et al., 2012). This is an immense national cost for the United States population and justifies the need to prioritize preventing melanoma.

Considering that melanoma is one of the most deadly skin cancers, the most extreme effect is death. On a large scale, death by melanoma can impact larger populations, especially considering how an ageing population and climate change can drive a greater rate of melanoma occurrences. In 2018, there were 287,723 cases of melanoma skin cancer, and approximately 60,712 people had died of melanoma skin cancer ("2020 Melanoma," 2020). This is a significant statistic and demonstrates the immense effect melanoma can have on a population. On a microscale, the effect of death can impact families and community members and perhaps the productivity that individuals contributed before death. However, as the rising trends of melanoma occur, the macroscale effects can drive changes to entire populations.

Also, it is essential to note that age plays a crucial role in melanoma as a skin disease. Specifically, old age contributes to an increased chance of melanoma and a greater vulnerability to the adverse effects of melanoma. According to a 2017 report by the United Nations, the global population aged 60 plus years in 1980 was 382 million; in 2017, the senior population rose to 962 million ("World Population," 2017). Based on these patterns, by 2050, the global senior population is expected to be 2.1 billion ("World Population," 2017). Clearly, the global population is rising, which could pose a significant problem for the future of skin cancer, more specifically melanoma. Firstly, the average age of people diagnosed with melanoma is 65, and chance increases with age. ("World Population," 2017). Also, the chance of melanoma occurrence increases with age, and the specific conditions and adverse effects can also worsen.

A study intended to determine the impact of melanoma on a population aged 65 and older completed a retrospective review of 481 patients with 525 primary melanomas (Rees et al., 2016). The study found that the survival rate is proportional with age; specifically, older patients presented with poorer prognosis melanomas and were less likely to receive adequate surgical excision margins resulting in higher rates of local recurrence (Rees et al., 2016). In melanoma patients of 65 years, the increasing number of medical comorbidities explains much of the age-related variations in overall survival and disease specific survival (Rees et al., 2016). Overall, the study emphasized that the senior population presenting with melanoma had a lesser chance of survival and a greater chance of having a varying combination of conditions that could ultimately contribute to a worse outcome for the patient. Considering that the ageing population is on the rise, the impact of melanoma on national, global and public health will not only continue to be immense, but it will also increase in its extent of implications. Therefore people need to take the necessary precautions and make preventative efforts to combat the greater vulnerability of the ageing global population.

Furthermore, one of the most significant environmental causes of melanoma is exposure to U.V. radiation. Specifically, a U.K. study found that 86% of melanomas are associated with exposure to the sun (Memon et al., 2021). Due to ozone depletion, the average member of the public has a greater exposure today to the sun and its harmful cancer-causing radiation compared to past generations. Specifically, an essay called "Impact of Climate Change on Skin Cancer" concluded that ozone depletion and climate change are separate entities that are intricately linked, and they both can increase the incidence of skin cancer through different means (Bharath & Turner, 2009). Considering that today both climate change and ozone depletion are major environmental issues that can ultimately lead to a tremendous increase in melanoma occurrences, it is clear that alongside an ageing population, the rate of melanoma will tend upward going into future generations. Thus, this immense disease will continue to wreak havoc on national, global, and public health.

In conclusion, although melanoma impacts begin with the individual, they transgress to have large-scale national, global, and public health repercussions. Specifically, large bodies of patients suffer such that they have a decreased quality of life and are less productive in their contributions to society due to their inflection with melanoma. Also, the economic repercussions can be very harmful to many people such as patients, family members, and taxpayers. In the most extreme and devastating scenarios, which are unfortunately common, death due to melanoma-related illness is an effect that can impact large populations of people. This death by melanoma not only results in significant emotional effects for those left behind but also implications that can affect national and even global productivity. The immense national and global effects emphasize the importance of prioritizing preventative efforts and awareness about melanoma and skin cancer in general. Through the increased awareness and investments in research, preventative and treatment methods, the world can better tackle this continually impactful global issue.

Additionally, it is essential not only to consider the large-scale repercussions of melanoma but also how the future, which entails a more senior population and UV-exposed planet, will ultimately make humankind much more vulnerable to the deadly wrath of melanoma. For that reason, people need to be extra careful in their preventative efforts against skin cancer, especially melanoma, and the general public and scientists need to be ready to make changes to the evolving situation of melanoma.

Lastly, it is essential to note that although melanoma is an immensely powerful and deadly form of skin cancer, it is primarily preventable. As previously mentioned, 86% of melanoma is associated with U.V.

radiation from sun exposure (Memon et al., 2021). Therefore, considering that it is highly preventable if more of the general public learn the importance of melanoma prevention and how to live a lifestyle that decreases exposure to melanoma-causing factors, the story of an increasing trend can easily change. Future generations can avoid the many national, global and public health repercussions associated with melanoma if preventative efforts start today.

Future Direction of Melanoma Detection, Prevention and Treatment

Massa Mohamed Ali

Melanoma is a highly aggressive form of uncontrolled cell growth that accounts for 60-80% of deaths from skin cancer. When it spreads deeper into the skin or other parts of the body, melanoma can be one of the most dangerous types of cancers (Sadozai et al., 2017). According to the book Possibilities for the Therapy of Melanoma: Current Knowledge and Future Directions, melanomas represent 3% of all skin cancers but 65% of skin cancer deaths (Valko-Rokytovská et al., 2017). In the past decade, researchers have overall made significant advancements in the field of cancer immunotherapy and in the detection, prevention, and treatment options available for malignancies, including several types of melanoma. But where is the melanoma research field going, and what should one expect in the future as researchers start to move beyond the standard cancer treatments? This chapter outlines the current and future melanoma research developments. It discusses the advancement of detection, prevention, and treatment methods, including the role of new technology and drug therapy.

Detection

Detecting melanoma early is important to prevent the cancer from spreading throughout the body, allow for more treatment options, and save lives. Current detection methods rely on skin self-exams and exams by health care professionals. For skin self-exams, knowing one's own skin is vital. One must routinely check for any abnormalities in their pattern of moles, blemishes, freckles, and other marks on the skin. Doctors recommend checking on one's own skin once a month, in a well-lit room, and using a full-length mirror. If new spots emerge or a change in size, shape, or colour of old spots occurs, a doctor must examine the patient as soon as possible. The exams done by health care professionals like doctors and dermatologists involve the use of dermoscopy, biopsy, and imaging techniques ("Can Melanoma Be Found Early?" n.d.). These detection methods are discussed in further detail in Chapter 3: "Diagnosing Melanoma" of this book.

A recent study done on 593 patients who were at a high risk for developing melanoma revealed that most melanomas were detected by clinicians using specialized imaging, and that less than 10% of the melanomas were detected by patients (Ferris, 2021). This suggests that self-examinations are not always sufficient for routine check-ups, especially for those who are at a high risk for developing melanoma. They need to visit doctors regularly and have them perform examinations and scans as needed. According to the National Cancer Institute, the "ABCDE" method is used to examine spots and moles on the skin. The "ABCDE" method stands for the following: "Asymmetry: The shape of one half of the spot does not match the other half. Border that is irregular: The edges are often ragged, notched, or blurred in outline. The pigment may spread into the surrounding skin. Color that is uneven: Shades of black, brown, and tan may be present. Areas of white, gray, red, pink, or blue may also be seen. Diameter: There is a change in size, usually an increase. Melanomas can be tiny, but most are larger than the size of a pea (larger than 6 millimeters or about 1/4 inch). Evolving: The mole has changed over the past few weeks or months" ("ABCDE's of Melanoma," n.d.). Currently, the most common way to diagnose melanoma is to perform a biopsy and remove the abnormal tissue to examine it for cancer cells ("What Does Melanoma Look Like?" n.d.). There are, however, newer approaches being developed, studied, and used that do not require the removal of a skin sample. These techniques include reflectance confocal microscopy (RCM) and adhesive patch testing. RCM is a US Food and Drug Administration–approved type of optical biopsy that does not require cutting into the skin. It uses novel technology that provides noninvasive, in vivo imaging of the skin in great detail. Several studies have shown that this technology significantly improves diagnostic accuracy and early detection of melanocytic and non-melanocytic skin cancers when compared with clinical and dermoscopic examinations alone. Even the combination of RCM imaging with dermoscopy has been shown to improve the accuracy of skin cancer diagnosis more than using dermoscopy alone (Levine & Markowitz, 2018; "What Does Melanoma Look Like?" n.d., Zhao & He, 2010). Using RCM has also reduced the number of biopsies of benign tumours and the number of unnecessary patient skin excisions. This technology decreases the health care expenses and the distress associated with the complete removal of skin spots. RCM can also determine the edges of a melanoma to help surgeons remove the accurate amount of malignant cell growth. Additionally, RCM imaging has been shown to help diagnose even the most challenging types of melanoma - lentigo maligna and lentigo maligna melanoma. According to an editorial by doctors Amanda Levine and Orit Markowitz in New York, these two subtypes "are often diagnostically challenging to both dermatologists and dermatopathologists because of ill-defined borders and overlapping features with solar lentigines, pigmented actinic keratosis, lichen planus–like keratosis, and seborrheic keratosis. These lesions are often large in diameter and/or located near cosmetically sensitive areas making an excisional biopsy impractical and unfeasible. Small incisional biopsies at the darkest and/or thickest areas of the lesion do not always correlate with the most advanced areas histologically, as there are significant histologic variability and skip features within a given lesion; therefore, limited sampling may be inadequate for diagnosis." With RCM, diagnosing lentigo maligna and lentigo maligna melanoma becomes easier and more accurate, as RCM can help select more specific areas for biopsies and reduce sampling error (Levine & Markowitz, 2018). According to the same editorial, "in a clinical setting, RCM imaging will benefit patients by providing same-day diagnosis of a cutaneous disease as well as confirmation of treatment efficacy, or as a perioperative tool to aid in cutaneous surgeries. RCM's ability to scan the entire lesion and noninvasively determine the most diagnostically and prognostically significant area to biopsy will help to reduce the risk of sampling error and false-negative rates owing to heterogeneity within lesions." Therefore, RCM imaging seems to be continually evolving to become more widespread and useful in the future. It is currently used in Europe and available in some centres of the United States and Canada (Levine & Markowitz, 2018). The second approach being studied as an alternative melanoma detection and diagnosis method is the adhesive patch testing technique. This method uses a pain-free adhesive patch that looks for the genes found in melanoma. The patch is placed over the suspicious area and is then removed. Once removed, the patch takes away some of the top layers of the skin with it, which are then tested for certain gene changes associated with cancerous skin cells. The results of this test are usually found in less than a week and they determine whether a biopsy is needed or not. If a gene change is found, doctors perform a standard biopsy of the area. If a gene change is not found, a biopsy is not needed and the suspicious area of the skin can be monitored instead ("Melanoma Skin Cancer Research," n.d.). There are also modern lab tests that can be performed to determine the appropriate treatment for each patient after being diagnosed with melanoma through RCM or patch testing. Lab tests like DecisionDx-Melanoma divide melanomas into two main groups based on their gene patterns. Researchers developed this in 2013 after they found that certain gene expression patterns in melanoma cells can help determine whether the melanoma is likely to spread or not. DecisionDx-Melanoma identifies two main groups: Class 1 tumours, which have a lower risk of spreading and Class 2 tumours, which have a higher risk of spreading. This information is useful when considering treatment options, because it can help identify if a patient with early-stage melanoma needs a sentinel lymph node biopsy (SLNB) (to determine whether cancer cells have spread to a nearby lymph node meaning stage III melanoma) or if a patient needs additional treatment, follow-up, or surgery ("Melanoma Skin Cancer Research," n.d.). According to a systematic review of the literature on DecisionDx-Melanoma published on the American Journal of Clinical Dermatology, DecisionDx-Melanoma has been tested on over 3900 patient samples and has shown to be consistently effective, accurate, and useful ("Castle Biosciences Announces Level of Evidence Review on DecisionDx-Melanoma Published in Latest Issue of American Journal of Clinical Dermatology," 2019). This area of research, along with the testing for other genes and gene patterns involved in the development of melanoma, is still being actively studied and more advances in technology like DecisionDx-Melanoma are expected in the near future.

Prevention

Another area that is currently being actively studied with ongoing clinical trials is the possibility of creating a vaccine for melanoma. The vaccine aims to stimulate the body's natural immune response to destroy cancerous cells in the body, similar to the way viral vaccines work. These vaccines, such as polio, measles, and mumps vaccines, usually contain weakened viruses or their parts and they cause the body to build and prepare an attack. However, treating and preventing a cancerous disease like melanoma is much more difficult than preventing viral infections. The results of the studies using vaccines to treat melanoma have been mixed ("What's New in Melanoma Skin Cancer Research?" n.d.). However, newer vaccines and prevention methods are now being studied and have shown to have some potential. For instance, since the environmental exposure to ultraviolet (UV) radiation in sunlight or tanning beds are strong risk factors for the genesis of skin cancer, there has been some recent research in the area of sun protection and in reducing indoor tanning practices as preventative measures. Doctors believe that UV radiation is the leading cause of melanoma, which indicates that the main melanoma prevention methods must include less exposure to UV radiation and more public awareness about its harm ("Advances in Melanoma and Other Skin Cancers Research," n.d.).

Treatment

Cancer treatment is known to rely on surgery, chemotherapy, and radiation therapy. However, during the past decade, significant changes have occurred in the treatment options available for skin cancer and other melanoma-related abnormalities. Due to the detailed knowledge of molecular biology, the immune response, and the genetic mechanisms behind melanoma, two newer treatment options became available - immunotherapy and targeted therapy. These treatment options are available mainly for more advanced stages of melanoma, as surgery remains the standard treatment for earlier stages of melanoma. A more detailed outline of the treatment options for the different stages of melanoma can be found in chapters 7 and 8 of this book. Moreover, both immunotherapy and targeted therapy have led to dramatic improvements in survival for patients with advanced melanoma. They are both constantly being researched and improved (Valko-Rokytovská et al., 2017). Immunotherapy helps the body's immune system attack melanoma cells more effectively. Since melanoma tends to have a relatively high number of genetic mutations, the immune system can recognize the mutations more easily than those of other cancer types. This makes a patient's immune system more likely to respond to immunotherapies and attack the easily recognized mutated cells. A type of immunotherapy, called immune checkpoint inhibition, uses immune checkpoint inhibitors (ICIs) to help the body attack melanoma cells. The three newer approved ICIs are ipilimumab (Yervoy), pembrolizumab (Keytruda), and nivolumab (Opdivo). Using these drugs as treatment has shown impressive results in people with advanced melanoma and has been approved for some patients with metastatic (the kind that spreads to other sites of the body through the lymphatic system and/or blood vessels) melanoma and unresectable (cannot be removed with surgery) melanoma ("What's New in Melanoma Skin Cancer Research?" n.d.). Interestingly, a clinical trial has also shown that a combination of ipilimumab and nivolumab can shrink the melanoma that has spread to the brain in some patients. The results show that "brain metastases shrank in half of the 94 participants with sufficient follow-up, including 24 participants whose brain metastases were completely eliminated and 28 whose brain metastases shrunk partially. Two other participants had brain metastases that remained stable — meaning, they did not shrink or grow — for at least 6 months." Some of the tumour reductions appeared rapidly and were detectable only 6 weeks after the start of the treatment. However, as part of an editorial, Drs. Samra Turajlic and James Larkin wrote that although the immunotherapy combination seems to be a promising potential treatment for people with metastatic melanoma tumors in the brain, one cannot assume that it will also benefit higher-risk patients who were excluded from the study; more research must be done ("Nivolumab and Ipilimumab Effective against Melanoma That Has Spread to the Brain," 2018). Researchers are actively testing combinations of the ICIs to find ways to increase the number of people who can benefit from immunotherapy treatments. Some researchers have combined ICIs with general immune stimulants to produce a kind of chemical alarm in the body to alert the immune system that a threat exists. A small clinical trial combined pembrolizumab with a specific type of general immune stimulant and found that melanoma tumours shrank in approximately 80% of people who received the two treatments together. There are currently larger trials of this study being conducted, along with tests for other combinations to try and treat as many patients as possible ("Advances in Melanoma and Other Skin Cancers Research," n.d.). Another type of immunotherapy, called adoptive cell therapy (ACT), is also being tested to treat patients with metastatic melanoma. ACT identifies the immune cells that recognize the tumour best in a patient. After the identification process, the desired cells are grown in a lab and given back to the patient to help fight the tumour cells. In early clinical trials, around 50% of the patients undergoing ACT saw some shrinking in their tumours. However, ACT is complicated, expensive, and may have dangerous side effects. Researchers are currently testing ways to make the immune cells collected from patients better at killing cancer cells before giving them back to the patient. They are also testing to find common proteins in cancer cells that are recognized by immune cells to make them target the cancer cells more efficiently ("Advances in

Melanoma and Other Skin Cancers Research," n.d.). An additional form of therapy that is currently being explored is adjuvant therapy. Adjuvant therapy is an extra cancer treatment given to patients after the primary treatment. The ICIs mentioned above have all been approved as adjuvant therapies for melanoma, including metastatic melanoma. According to the United States' National Cancer Institute, "in clinical trials, all three immune checkpoint inhibitors reduced the risk of recurrence for some patients when given after surgery, although serious side effects were seen in many trial participants. A follow-up clinical trial is now looking at whether some patients with earlier stage melanoma at high risk of recurrence may benefit from ICI therapy after surgery." Not only are researchers studying the effects of adjuvant therapy given after primary treatment (such as surgery), but they are also exploring immunotherapy options given before the primary treatment. A current study is comparing patients who receive pembrolizumab both before and after surgery with those who receive the drug only after surgery. The results of this study, among others, will provide more evidence about the effectiveness of using immunotherapy before and after primary treatment ("Advances in Melanoma and Other Skin Cancers Research," n.d.). Moreover, aside from immunotherapy, targeted therapy is another treatment option for patients with melanoma. Targeted therapy uses drugs to focus on parts of the melanoma cells that make them different from normal cells. This treatment is different from standard chemotherapy, as its side effects are not the same and can be less severe. The drugs currently being studied for targeted therapy mainly target the BRAF gene, MEK proteins, the C-KIT gene, and other gene or protein changes. The BRAF gene changes are responsible for about half of all melanoma cells, because mutations in BRAF help the cells grow. The American Cancer Society notes that the drugs that target the BRAF protein or the related MEK proteins have been shown to shrink many of these tumors, especially when BRAF and MEK inhibitors are combined. These drugs are now used to treat advanced melanomas that test positive for the BRAF gene change. Researchers are currently looking at whether these drugs might be helpful before or after surgery for some earlier stage melanomas. The American Cancer Society also mentions that drugs like imatinib (Gleevec), dasatinib (Sprycel), and nilotinib (Tasigna), are being tested in clinical trials for their effectiveness in targeting the changes in the C-KIT gene. There are also other similar drugs being studied as well. Some examples include axitinib (Inlyta), pazopanib (Votrient), and everolimus (Afinitor). There are also ongoing studies about the possibility of combining these targeted drugs with other types of treatment like chemotherapy or immunotherapy ("What's New in Melanoma Skin Cancer Research?" n.d.). Overall, the treatment options available for advanced stage melanoma are being studied and improved on a daily basis. The treatments were first revolutionized in 2011 and have continued improving steadily ever since. In 2011, ipilimumab and vemurafenib, the two ICIs that improved overall survival in phase III clinical trials, were approved for use in the United States. Prior to 2011, there was no effective, non-toxic treatment for melanoma. Now, there are multiple different approved treatment options, including the more recent aforementioned immunotherapy and targeted therapy. The survival of patients has improved to about 4 to 5 years and there are patients with metastatic melanoma who were even cured. However, with all these advancements, there is still more that needs to be done to improve the treatment options available for patients with melanoma (Shah & VanderWalde, 2019). At the West Cancer Centre and Research Institute, Dr. Ari VanderWalde believes that researchers need to focus on the treatment options available for earlier stage melanoma. Researchers have not yet found a way to be able to integrate potentially useful therapies into earlier stage melanoma, especially stage II with immunotherapy agents. In an interview, Dr. VanderWalde said that those with stage II disease can actually have a poorer long-term outcome than patients with stage IIIA, who have more effective treatment options (Shah & VanderWalde, 2019). Whether it be through different combinations of ICIs, or advanced immunotherapy techniques, or changing the timing of the drugs given, most melanoma researchers are looking to understand how best to use existing therapies. There are still questions about various aspects of the melanoma treatment process. Researchers are studying whether patients with BRAF mutations should receive targeted drugs or ICIs first. They are looking for biomarkers in melanoma that can predict which tumors might respond to available immunotherapies. They are also exploring the optimal times and amounts of the drugs given to ensure that the development of resistance to drugs is delayed and to make the drugs generally easier to tolerate without risking the recurrence of melanoma ("Advances in Melanoma and Other Skin Cancers Research," n.d.). While there seems to be many questions left unanswered, there are also many active clinical trials and studies being published constantly. Overall, the field of skin cancer research involving the detection, prevention, and treatment of melanoma is promising, has a lot of potential and is increasingly helping many people around the world.

References

Chapter 1:

Berwick, M., Buller, D. B., Cust, A., Gallagher, R., Lee, T. K., Meyskens, F., Pandey, S., Thomas, N. E., Veierød, M. B., & Ward, S. (2015). Melanoma Epidemiology and Prevention. *Melanoma, 17*–49. https://doi.org/10.1007/978-3-319-22539-5_2

Bradford, P. T., Freedman, D. M., Goldstein, A. M., & Tucker, M. A. (2010). Increased Risk of Second Primary Cancers After a Diagnosis of Melanoma. *Archives of Dermatology, 146*(3). https://doi.org/10.1001/archdermatol.2010.2

Faries, M. B., & Curti, B. D. (2018). Melanoma. *Oral, Head and Neck Oncology and Reconstructive Surgery,* 876–891. https://doi.org/10.1016/b978-0-323-26568-3.00043-9

Lee, C., Collichio, F., Ollila, D., & Moschos, S. (2013). Historical review of melanoma treatment and outcomes. *Clinics in Dermatology, 31*(2), 141–147. https://doi.org/10.1016/j.clindermatol.2012.08.015

Melanoma Treatment (PDQ®)–Health Professional Version.
(2021, February 5). National Cancer Institute; Cancer.
gov. https://www.cancer.gov/types/skin/hp/
melanoma-treatment-pdq#_1

Mitchell, T. C., Karakousis, G., & Schuchter, L. (2020).
Melanoma. *Abeloff's Clinical Oncology*, 1034-1051.e2.
https://doi.org/10.1016/b978-0-323-47674-4.00066-9

Rebecca, V. W., Sondak, V. K., & Smalley, K. S.
M. (2012). A brief history of melanoma. *Melanoma
Research, 22*(2), 114–122. https://doi.org/10.1097/
cmr.0b013e328351fa4d

World Health Organization: WHO. (2019, July 12).
Cancer. Who.int; World Health Organization: WHO.
https://www.who.int/health-topics/cancer#tab=tab_1

Chapter 2:

Fayed, L. (2020). *When was cancer first discovered?*
https://www.verywellhealth.com/the-history-of-cancer-514101

Papavramidou N, Papavramidis T, Demetriou T. Ancient Greek and Greco-Roman methods in modern surgical treatment of cancer. *Ann Surg Oncol.* 2010;17(3):665–667. doi:10.1245/s10434-009-0886-6

Faguet GB. A brief history of cancer: age-old milestones underlying our current knowledge database. *Int J Cancer.* 2015;136(9):2022-36. doi:10.1002/ijc.29134
Early History of Cancer. American Cancer Society. (n.d.). https://www.cancer.org/cancer/cancer-basics/history-of-cancer/what-is-cancer.html

American Cancer Society. Early theories about cancer causes. Updated June 12, 2014
Di Lonardo A, Nasi S, Pulciani S. Cancer: we should not forget the past. *J Cancer.* 2015;6(1):29–39. Published 2015 Jan 1. doi:10.7150/jca.10336

Walter E, Scott M. The life and work of Rudolf Virchow 1821-1902: "Cell theory, thrombosis and the sausage duel". *J Intensive Care Soc.* 2017;18(3):234–235. doi:10.1177/1751143716663967

Weiss L. Early concepts of cancer. *Cancer Metastasis Rev.* 2000;19:205–17

Hippocrate. Prorrhétique II. In: Littré E, editor. Oeuvres complètes d'Hippocrate. Vol. 9. Paris: J. B. Baillière; 1861. p. 32.

Hippocrate. Aphorismes VI. In: Littré E, editor. Oeuvres complètes d'Hippocrate. Vol. 4. Paris: J. B. Baillière; 1844. p. 572.

Daremberg Ch. Oeuvres d'Oribase. Vol. 4. Paris: Imprimerie Nationale; 1862. p. 244.

Strauss, M. (2021, May 3). *Earliest Human Cancer Found in 1.7-Million-Year-Old Bone.* Culture. https://www.nationalgeographic.com/culture/article/oldest-human-cancer-disease-origins-tumor-fossil-science?loggedin=true.

Encyclopædia Britannica, inc. (n.d.). *Sir Percivall Pott.* Encyclopædia Britannica. https://www.britannica.com/biography/Percivall-Pott.

Kirkpatrick, D. B. (1984, November 1). *The first primary brain-tumor operation.* jns. https://thejns.org/view/journals/j-neurosurg/61/5/article-p809.xml.

Who Discovered Lung Cancer. Newswire. (2013, October 6). https://www.newswire.com/who-discovered-lung-cancer/243157.

WHO. Skin cancers. Available at: http://www.who.int/uv/faq/skincancer/en/index1.html; Last accessed April 2015.

Roguin A. Rene Theophile Hyacinthe Laënnec (1781-1826): the man behind the stethoscope. Clin Med Res. 2006;4:230-5.

Chapter 3:

American Cancer Society. (2019). *Tests For Melanoma Skin Cancer | Melanoma Diagnosis*. Cancer.org. Retrieved 14 May 2021, from https://www.cancer.org/cancer/melanoma-skin-cancer/detection-diagnosis-staging/how-diagnosed.html.

Bauman, C. A., Emary, P., Damen, T., & Dixon, H. (2018). Melanoma in situ: a case report from the patient's perspective. *The Journal of the Canadian Chiropractic Association, 62*(1), 56–61.

Blundo, A., Cignoni, A., Banfi, T., & Ciuti, G. (2021). Comparative Analysis of Diagnostic Techniques for Melanoma Detection: A Systematic Review of Diagnostic Test Accuracy Studies and Meta-Analysis. *Frontiers In Medicine, 8*. https://doi.org/10.3389/fmed.2021.637069

Canadian Cancer Society. *Diagnosis of melanoma skin cancer*. www.cancer.ca. Retrieved 13 May 2021, from https://www.cancer.ca/en/cancer-information/cancer-type/skin-melanoma/staging/?region=on.

Canadian Cancer Society. *Stages of melanoma skin cancer - Canadian Cancer Society*. www.cancer.ca. Retrieved 13 May 2021, from https://www.cancer.ca/en/cancer-information/cancer-type/skin-melanoma/diagnosis/?region=on

Canadian Dermatology Association. (2021). *Melanoma - Canadian Dermatology Association*. Canadian Dermatology Association. Retrieved 13 May 2021, from https://dermatology.ca/public-patients/skin/melanoma/.

CDC. (2021). *What Are the Symptoms of Skin Cancer? | CDC*. Cdc.gov. Retrieved 13 May 2021, from https://www.cdc.gov/cancer/skin/basic_info/symptoms.htm.

Curiel-Lewandrowski, C., Novoa, R., Berry, E., Celebi, M., Codella, N., & Giuste, F. et al. (2019). Artificial Intelligence Approach in Melanoma. *Melanoma*, 1-31. https://doi.org/10.1007/978-1-4614-7322-0_43-1

Davis, L., Shalin, S., & Tackett, A. (2019). Current state of melanoma diagnosis and treatment. *Cancer Biology & Therapy, 20*(11), 1366-1379. https://doi.org/10.1080/15384047.2019.1640032

Ferlay, J., Colombet, M., Soerjomataram, I., Parkin, D., Piñeros, M., Znaor, A., & Bray, F. (2021). Cancer statistics for the year 2020: An overview. *International Journal Of Cancer.* https://doi.org/10.1002/ijc.33588

Fried, L., Tan, A., Bajaj, S., Liebman, T., Polsky, D., & Stein, J. (2020). Technological advances for the detection of melanoma. *Journal Of The American Academy Of Dermatology, 83*(4), 996-1004. https://doi.org/10.1016/j.jaad.2020.03.122

Holmes, G. A., Vassantachart, J. M., Limone, B. A., Zumwalt, M., Hirokane, J., & Jacob, S. E. (2018). Using Dermoscopy to Identify Melanoma and Improve Diagnostic Discrimination. *Federal practitioner : for the health care professionals of the VA, DoD, and PHS, 35*(Suppl 4), S39–S45

Lee, H., & Kwon, K. (2019). Diagnostic techniques for improved segmentation, feature extraction, and classification of malignant melanoma. *Biomedical Engineering Letters, 10*(1), 171-179. https://doi.org/10.1007/s13534-019-00142-8

Leiter, U., Keim, U., & Garbe, C. (2020). Epidemiology of Skin Cancer: Update 2019. *Sunlight, Vitamin D And Skin Cancer,* 123-139. https://doi.org/10.1007/978-3-030-46227-7_6

Matthews, N., Li, W., Qureshi, A., Weinstock, M., & Cho, E. (2017). Epidemiology of Melanoma. *Cutaneous Melanoma: Etiology And Therapy,* 3-22. https://doi.org/10.15586/codon.cutaneousmelanoma.2017.ch1

Naik, P. (2021). Cutaneous Malignant Melanoma: A Review of Early Diagnosis and Management. *World Journal Of Oncology, 12*(1), 7-19. https://doi.org/10.14740/wjon1349

Pluta, R. (2011). Melanoma. *JAMA, 305*(22), 2368. https://doi.org/10.1001/jama.2011.658

Rastrelli, M., Tropea, S., Rossi, C., & Alaibac, M. (2014). *Melanoma: epidemiology, risk factors, pathogenesis, diagnosis and classification.* PubMed. Retrieved 12 May 2021, from https://pubmed.ncbi.nlm.nih.gov/25398793/.

Rossi, M., Pellegrini, C., Cardelli, L., Ciciarelli, V., Di Nardo, L., & Fargnoli, M. C. (2019). Familial Melanoma: Diagnostic and Management Implications. *Dermatology practical & conceptual, 9*(1), 10–16. https://doi.org/10.5826/dpc.0901a03

Stanford Health Education. (2020). *Genetic Assessment and Melanoma.* Stanfordhealthcare.org. Retrieved 13 May 2021, from https://stanfordhealthcare.org/medical-conditions/cancer/melanoma/assets/genetic-assessment-for-melanoma.html.

Swetter, S., Pollitt, R., Johnson, T., Brooks, D., & Geller, A. (2011). Behavioral determinants of successful early melanoma detection. *Cancer, 118*(15), 3725-3734. https://doi.org/10.1002/cncr.26707

Swetter, S., Tsao, H., Bichakjian, C., Curiel-Lewandrowski, C., Elder, D., & Gershenwald, J. et al. (2018). Guidelines of care for the management of primary cutaneous melanoma. *Journal Of The American Academy Of Dermatology, 80*(1), 208-250. https://doi.org/10.1016/j.jaad.2018.08.055

Tripp, M., Watson, M., Balk, S., Swetter, S., & Gershenwald, J. (2016). State of the science on prevention and screening to reduce melanoma incidence and mortality: The time is now. *CA: A Cancer Journal For Clinicians, 66*(6), 460-480. https://doi.org/10.3322/caac.21352

Walter, F. M., Prevost, A. T., Vasconcelos, J., Hall, P. N., Burrows, N. P., Morris, H. C., Kinmonth, A. L., & Emery, J. D. (2013). Using the 7-point checklist as a diagnostic aid for pigmented skin lesions in general practice: a diagnostic validation study. *The British journal of general practice : the journal of the Royal College of General Practitioners, 63*(610), e345–e353. https://doi.org/10.3399/bjgp13X667213
Weinstein, D., Leininger, J., Hamby, C., & Safai, B. (2014). Diagnostic and prognostic biomarkers in melanoma. *The Journal of clinical and aesthetic dermatology, 7*(6), 13–24.

Chapter 4:

Akhilanda, L. (n.d.). *Oncology Social Worker Q&A: Supporting Cancer Patients and Hospitals - The Patient Story*. The Patient Story | For Cancer Patients & Caregivers. https://www.thepatientstory.com/caregivers/lia-akhilanda/.

American Academy of Dermatology Association. (n.d.). *Financial help for people who have skin cancer*. American Academy of Dermatology. https://www.aad.org/public/diseases/skin-cancer/types/common/melanoma/financial-help.

American Cancer Society. (2019). *Short and Simple Melanoma Skin Cancer Guide: Understanding Melanoma*. American Cancer Society. https://www.cancer.org/cancer/melanoma-skin-cancer/if-you-have-melanoma.html.

American Cancer Society. (2020). *Surviving Melanoma Skin Cancer: Melanoma Survivor*. American Cancer Society. https://www.cancer.org/cancer/melanoma-skin-cancer/after-treatment/follow-up.html.

Cancer Care. (2021). *Melanoma Patient Support Group*. CancerCare. https://www.cancercare.org/support_groups/132-melanoma_patient_support_group.

Cancer.Net. (2020, September 2). *Melanoma - Coping with Treatment.* Cancer.Net. https://www. cancer.net/cancer-types/melanoma/coping-with-treatment#:~:text=You%20can%20have%20 emotional%20and,feel%20to%20their%20loved%20ones.

Cheung, W. Y., Bayliss, M. S., White, M. K., Stroupe, A., Lovley, A., King-Kallimanis, B. L., & Lasch, K. (2018). Humanistic burden of disease for patients with advanced melanoma in Canada. *Supportive Care in Cancer, 26*(6), 1985–1991. https://doi.org/10.1007/ s00520-017-4025-9

Government of British Columbia. (2016, January 8). *Protecting Yourself from Ultraviolet (UV) Radiation.* Province of British Columbia. https://www2.gov. bc.ca/gov/content/health/keeping-bc-healthy-safe/ radiation/ultraviolet-uv-radiation/protecting-yourself-from-ultraviolet-uv-radiation.

MATW. (2020). *Ali's Story.* MATW Project. https:// matwproject.org/about-ali/.

Melanoma Network of Canada. (2020, December 17). *Managing Life with Melanoma.* Melanoma Network of Canada. https://www.melanomanetwork.ca/ managingmelanoma/.

Melanoma Network of Canada. (2021, April 15). *Melanoma Support Groups.* Melanoma Network of Canada. https://www.melanomanetwork.ca/ supportgroups/.

Rao, M., Afshin, A., Singh, G., & Mozaffarian, D. (2013). Do healthier foods and diet patterns cost more than less healthy options? A systematic review and meta-analysis. *BMJ Open, 3*(12). https://doi.org/10.1136/bmjopen-2013-004277

Smart Patients. (n.d.). Smart Patients. https://www.smartpatients.com/communities/melanoma.

Team, G. T. E. (2018, June 11). *Mental Health Support for Cancer.* Psychotherapy for Cancer, Counseling for Cancer, Therapist for Cancer –. https://www.goodtherapy.org/learn-about-therapy/issues/cancer/support.

Ziembicki. (2021). *What to Eat & Drink During Melanoma Treatment // .* What to Eat & Drink During Melanoma Treatment. https://www.uhn.ca/PrincessMargaret/PatientsFamilies/Patient_Family_Library/diet_nutrition/pages/eat_drink_melanoma_treatment.aspx.

Chapter 5:

Heistein, J, B., & Acharya, U. (2020). *Malignant Melanoma*. StatPearls Publishing. https://www.ncbi.nlm.nih.gov/books/ NBK470409/#:~:text=A%20melanoma%20is%20a%20 tumor,the%20gastrointestinal%20tract%20and%20brain. *What Is Melanoma Skin Cancer?: What Is Melanoma?* American Cancer Society. (n.d.). https://www.cancer.org/cancer/melanoma- skin-cancer/about/what-is-melanoma. html#:~:text=Melanoma%20is%20a%20type%20 of,other%20areas%20of%20the%20body.

Curiel-Lewandrowski, C., Chen, S. C., & Swetter, S. M. (2012). Screening and Prevention Measures for Melanoma: Is There a Survival Advantage? *Current Oncology Reports, 14*(5), 458–467. https://doi.org/10.1007/s11912-012-0256-6

Rigel, D. S., & Carucci, J. A. (2000). Malignant melanoma: prevention, early detection, and treatment in the 21st century. *CA: A Cancer Journal for Clinicians, 50*(4), 215–236. https://doi.org/10.3322/ canjclin.50.4.215

Sander, M., Sander, M., Burbidge, T., & Beecker, J. (2020). The efficacy and safety of sunscreen use for the prevention of skin cancer. *Canadian Medical Association Journal, 192*(50). https://doi.org/10.1503/

cmaj.201085

Wehner, M. R. (2018). Sunscreen and melanoma prevention: evidence and expectations. *British Journal of Dermatology, 178*(1), 15–16. https://doi.org/10.1111/bjd.16111

Green, A. C., Williams, G. M., Logan, V., & Strutton, G. M. (2011). Reduced Melanoma After Regular Sunscreen Use: Randomized Trial Follow-Up. *Journal of Clinical Oncology, 29*(3), 257–263. https://doi.org/10.1200/jco.2010.28.7078

Protective *Clothing*. Melanoma Research Alliance. (n.d.). https://www.curemelanoma.org/about-melanoma/prevention/covering-up-with-clothing/.

Le Clair, M. Z., & Cockburn, M. G. (2016). Tanning bed use and melanoma: Establishing risk and improving prevention interventions. *Preventive Medicine Reports, 3,* 139–144. https://doi.org/10.1016/j.pmedr.2015.11.016

World Health Organization. (2017, July 27). *More can be done to restrict sunbeds to prevent increasing rates of skin cancer.* World Health Organization. https://www.who.int/phe/news/sunbeds-skin-cancer/en/#:~:text=Research%20shows%20that%20people%20who,of%20developing%20melanoma%20by%2059%25.

Ombra, M. N., Paliogiannis, P., Stucci, L. S., Colombino, M., Casula, M., Sini, M. C., Manca, A., Palomba, G., Stanganelli, I., Mandala, M., Gandini, S., Lissia, A., Doneddu, V., Cossu, A., Palmieri, G. (2019). Dietary compounds and cutaneous malignant melanoma: recent advances from a biological perspective. *Nutrition & Metabolism, 16*(1). https://doi.org/10.1186/s12986-019-0365-4

Mayo Foundation for Medical Education and Research. (2020, March 10). *Melanoma.* Mayo Clinic. https://www.mayoclinic.org/diseases-conditions/melanoma/symptoms-causes/syc-20374884.

Chapter 6:

American Cancer Society. (2021). *What Causes Melanoma? | Causes of Melanoma Skin Cancer*. Cancer. org. Retrieved 17 May 2021, from https://www.cancer. org/cancer/melanoma-skin-cancer/causes-risks-prevention/what-causes.html.

Cole, G. (2020). *Melanoma: Treatment, Causes, Types, Early Signs & Symptoms*. MedicineNet. Retrieved 17 May 2021, from https://www.medicinenet.com/melanoma/ article.htm.

Emanuel, P., & Cheng, H. (2013). *Melanoma pathology | DermNet NZ*. DermNet NA all about the skin. Retrieved 17 May 2021, from https://dermnetnz.org/ topics/melanoma-pathology/.

Hasney, C., Butcher, R., & Amedee, R. (2008). Malignant Melanoma of the Head and Neck: A Brief Review of Pathophysiology, Current Staging, and Management. *The Ochsner Journal*, 8(4), 181-185. Retrieved 17 May 2021, from.

Heistein, J., & Acharya, U. (2020). *Malignant Melanoma*. PubMed. Retrieved 17 May 2021, from https://www. ncbi.nlm.nih.gov/pubmed/29262210.

Melanoma Network of Canada. (2021). *Melanoma Stats and Facts*. Melanoma Network of Canada. Retrieved 17

May 2021, from https://www.melanomanetwork.ca/
stats-and-facts/.

physiopedia. (2021). *Malignant Melanoma.* Physiopedia.
Retrieved 17 May 2021, from https://www.physio-
pedia.com/Malignant_Melanoma#cite_note-DD-4.

Puckett, Y., Wilson, A., Farci, F., & Thevenin, C. (2020).
Melanoma Pathology. PubMed. Retrieved 17 May 2021,
from https://pubmed.ncbi.nlm.nih.gov/29083592/.

Swetter, S. (2019). *Cutaneous Melanoma: Background,
Pathophysiology, Etiology.* Emedicine.medscape.com.
Retrieved 17 May 2021, from https://emedicine.
medscape.com/article/1100753-overview#a4.

Whiteman, D., Watt, P., Purdie, D., Hughes, M.,
Hayward, N., & Green, A. (2003). Melanocytic Nevi,
Solar Keratoses, and Divergent Pathways to Cutaneous
Melanoma. *Journal Of The National Cancer Institute,
95*(11), 806-812. https://doi.org/10.1093/jnci/95.11.806

Chapter 7:

Canadian Cancer Society. (n.d.). *Chemotherapy for melanoma skin cancer – Canadian Cancer Society.* e/skin-melanoma/treatment/chemotherapy/?region=on Canadian Cancer Society. (n.d.). *Immunotherapy for melanoma skin cancer – Canadian Cancer Society.*

Canadian Cancer Society. (n.d.). *Radiation therapy for melanoma skin cancer – Canadian Cancer Society.*

Canadian Cancer Society. (n.d.). *Surgery for melanoma skin cancer – Canadian Cancer Society.*

Canadian Cancer Society. (n.d.). *Targeted therapy for melanoma skin cancer – Canadian Cancer Society.*

Dabrafenib plus Trametinib for advanced melanoma. (n.d.). https://www.cancer.gov/types/skin/research/dabrafenib-trametinib

Domingues, B., Lopes, J. M., Soares, P., & Pópulo, H. (2018). Melanoma treatment in review. *ImmunoTargets and therapy, 7,* 35–49.

Melanoma - Types of Treatment. (2020) https://www.cancer.net/cancer-types/melanoma/types-treatment

Smalley, K. S., Eroglu, Z., & Sondak, V. K. (2016). Combination Therapies for Melanoma: A New Standard of Care?. *American journal of clinical dermatology, 17*(2), 99–105.

Wróbel, S., Przybyło, M., & Stępień, E. (2019). The Clinical Trial Landscape for Melanoma Therapies. Journal of clinical medicine, *8(3)*, 368.

Chapter 8:

American Cancer Society. (2019, August 14). *Radiation Therapy for Melanoma | Melanoma Skin Cancer Radiation.* https://www.cancer.org/cancer/melanoma-skin-cancer/treating/radiation-therapy.html

American Cancer Society. (2019, August 14). *Surgery For Melanoma Skin Cancer | Melanoma Surgery Options.* https://www.cancer.org/cancer/melanoma-skin-cancer/treating/surgery.html

Atkins, M. B., Kunkel, L., Sznol, M., & Rosenberg, S. A. (2000). High-dose recombinant interleukin-2 therapy in patients with metastatic melanoma: Long-term survival update. *The Cancer Journal from Scientific American, 6 Suppl 1,* S11-14.

Atkins, M. B., Lotze, M. T., Dutcher, J. P., Fisher, R. I., Weiss, G., Margolin, K., Abrams, J., Sznol, M., Parkinson, D., Hawkins, M., Paradise, C., Kunkel, L., & Rosenberg, S. A. (1999). High-dose recombinant interleukin 2 therapy for patients with metastatic melanoma: Analysis of 270 patients treated between 1985 and 1993. *Journal of Clinical Oncology: Official Journal of the American Society of Clinical Oncology, 17*(7), 2105–2116. https://doi.org/10.1200/JCO.1999.17.7.2105

Batus, M., Waheed, S., Ruby, C., Petersen, L., Bines, S. D., & Kaufman, H. L. (2013). Optimal Management of

Metastatic Melanoma: Current Strategies and Future Directions. *American Journal of Clinical Dermatology, 14*(3), 179–194. https://doi.org/10.1007/s40257-013-0025-9

Bhatia, S., Tykodi, S. S., & Thompson, J. A. (2009). Treatment of metastatic melanoma: An overview. *Oncology (Williston Park, N.Y.), 23*(6), 488–496. Blesa, J. M. G., Enrique Grande Pulido, Vicente Alberola Candel, & Mariano Provencio Pulla. (2011). Melanoma: From Darkness to Promise: American Journal of Clinical Oncology. 34(2), 179–187.

Byrne, E. H., & Fisher, D. E. (2017). Immune and molecular correlates in melanoma treated with immune checkpoint blockade. *Cancer, 123*(S11), 2143–2153. https://doi.org/10.1002/cncr.30444

Canadian Cancer Society. (n.d.-a). *Chemotherapy for melanoma skin cancer – Canadian Cancer Society.* Www. Cancer.Ca. Retrieved May 15, 2021, from https://www.cancer.ca:443/en/cancer-information/cancer-type/skin-melanoma/treatment/chemotherapy/?region=on

Canadian Cancer Society. (n.d.-b). *Immunotherapy for melanoma skin cancer – Canadian Cancer Society.* Www.Cancer.Ca. Retrieved May 15, 2021, from https://www.cancer.ca:443/en/cancer-information/cancer-type/skin-melanoma/treatment/immunotherapy/?region=on

Canadian Cancer Society. (n.d.-c). *Radiation therapy for melanoma skin cancer – Canadian Cancer Society.* Www.Cancer.Ca. Retrieved May 15, 2021, from https://www.cancer.ca:443/en/cancer-information/cancer-type/skin-melanoma/treatment/radiation-therapy/?region=on

Canadian Cancer Society. (n.d.). *Surgery for melanoma skin cancer – Canadian Cancer Society.* Www.Cancer.Ca. Retrieved May 15, 2021, from https://www.cancer.ca:443/en/cancer-information/cancer-type/skin-melanoma/treatment/surgery/?region=on

Canadian Cancer Society. (n.d.-d). *Targeted therapy for melanoma skin cancer – Canadian Cancer Society.* Www.Cancer.Ca. Retrieved May 15, 2021, from https://www.cancer.ca:443/en/cancer-information/cancer-type/skin-melanoma/treatment/targeted-therapy/?region=on

Chapman, P. B., Hauschild, A., Robert, C., Haanen, J. B., Ascierto, P., Larkin, J., Dummer, R., Garbe, C., Testori, A., Maio, M., Hogg, D., Lorigan, P., Lebbe, C., Jouary, T., Schadendorf, D., Ribas, A., O'Day, S. J., Sosman, J. A., Kirkwood, J. M., … McArthur, G. A. (2011). Improved Survival with Vemurafenib in Melanoma with BRAF V600E Mutation. *New England Journal of Medicine, 364*(26), 2507–2516. https://doi.org/10.1056/NEJMoa1103782

Chiarion Sileni, V., Nortilli, R., Aversa, S. M., Paccagnella, A., Medici, M., Corti, L., Favaretto, A.

G., Cetto, G. L., & Monfardini, S. (2001). Phase II randomized study of dacarbazine, carmustine, cisplatin and tamoxifen versus dacarbazine alone in advanced melanoma patients. *Melanoma Research, 11*(2), 189–196. https://doi.org/10.1097/00008390-200104000-00015

Davis, L. E., Shalin, S. C., & Tackett, A. J. (2019). Current state of melanoma diagnosis and treatment. *Cancer Biology & Therapy, 20*(11), 1366–1379. https://doi.org/10 .1080/15384047.2019.1640032

DeVita Jr., V. T., & Chu, E. (2008). *A History of Cancer Chemotherapy.* 68(21). https://cancerres.aacrjournals. org/content/68/21/8643.short

Domingues, B., Lopes, J. M., Soares, P., & Pópulo, H. (2018). Melanoma treatment in review. *ImmunoTargets and Therapy*, 7, 35–49. https://doi.org/10.2147/ITT. S134842

Flaherty, K. T., Robert, C., Hersey, P., Nathan, P., Garbe, C., Milhem, M., Demidov, L. V., Hassel, J. C., Rutkowski, P., Mohr, P., Dummer, R., Trefzer, U., Larkin, J. M. G., Utikal, J., Dreno, B., Nyakas, M., Middleton, M. R., Becker, J. C., Casey, M., … METRIC Study Group. (2012). Improved survival with MEK inhibition in BRAF-mutated melanoma. *The New England Journal of Medicine, 367*(2), 107–114. https://doi. org/10.1056/NEJMoa1203421

George, D. D., Armenio, V. A., & Katz, S. C. (2017). Combinatorial immunotherapy for melanoma. *Cancer Gene Therapy, 24*(3), 141–147. https://doi.org/10.1038/cgt.2016.56

González, N., & Ratner, D. (2016). *Novel Melanoma Therapies and Their Side Effects. 97*(6), 426–428.

Hancock, B., Wheatley, K., Harris, S., Ives, N., Buck, G., Horsman, J., Middleton, M., Thatcher, N., Lorigan, P., Marsden, J., Burrows, L., & Gore, M. (2004). Adjuvant Interferon in High-Risk Melanoma: The AIM HIGH Study — United Kingdom Coordinating Committee on Cancer Research Randomized Study of Adjuvant Low-Dose Extended-Duration Interferon Alfa-2a in High-Risk Resected Malignant Melanoma. *Journal of Clinical Oncology : Official Journal of the American Society of Clinical Oncology, 22*, 53–61. https://doi.org/10.1200/JCO.2004.03.185

Harris, J. M., & Chess, R. B. (2003). Effect of pegylation on pharmaceuticals. *Nature Reviews Drug Discovery, 2*(3), 214–221. https://doi.org/10.1038/nrd1033

Hauschild, A., Grob, J.-J., Demidov, L. V., Jouary, T., Gutzmer, R., Millward, M., Rutkowski, P., Blank, C. U., Miller, W. H., Kaempgen, E., Martín-Algarra, S., Karaszewska, B., Mauch, C., Chiarion-Sileni, V., Martin, A.-M., Swann, S., Haney, P., Mirakhur, B., Guckert, M. E., … Chapman, P. B. (2012). Dabrafenib in BRAF-mutated metastatic melanoma: A multicentre, open-label, phase 3 randomised controlled trial. *Lancet*

(London, England), 380(9839), 358–365. https://doi.
org/10.1016/S0140-6736(12)60868-X

Hodi, F. S., O'Day, S. J., McDermott, D. F., Weber, R.
W., Sosman, J. A., Haanen, J. B., Gonzalez, R., Robert,
C., Schadendorf, D., Hassel, J. C., Akerley, W., van den
Eertwegh, A. J. M., Lutzky, J., Lorigan, P., Vaubel, J.
M., Linette, G. P., Hogg, D., Ottensmeier, C. H., Lebbé,
C., … Urba, W. J. (2010). Improved survival with
ipilimumab in patients with metastatic melanoma.
The New England Journal of Medicine, 363(8), 711–723.
https://doi.org/10.1056/NEJMoa1003466

Hoeflich, K. P., Merchant, M., Orr, C., Chan, J., Den
Otter, D., Berry, L., Kasman, I., Koeppen, H., Rice, K.,
Yang, N.-Y., Engst, S., Johnston, S., Friedman, L. S.,
& Belvin, M. (2012). Intermittent administration of
MEK inhibitor GDC-0973 plus PI3K inhibitor GDC-
0941 triggers robust apoptosis and tumor growth
inhibition. *Cancer Research, 72*(1), 210–219. https://doi.
org/10.1158/0008-5472.CAN-11-1515

Lee, C., Collichio, F., Ollila, D., & Moschos, S. (2013).
Historical review of melanoma treatment and outcomes.
Clinics in Dermatology, 31(2), 141–147. https://doi.
org/10.1016/j.clindermatol.2012.08.015

Liu, P., Zhang, C., Chen, J., Zhang, R., Ren, J., Huang,
Y., Zhu, F., Li, Z., & Wu, G. (2011). Combinational
therapy of interferon-α and chemotherapy normalizes
tumor vasculature by regulating pericytes including
the novel marker RGS5 in melanoma. *Journal of*

Immunotherapy (Hagerstown, Md.: 1997), 34(3), 320–326.
https://doi.org/10.1097/CJI.0b013e318213cd12

Livingstone, E., Zimmer, L., Vaubel, J., & Schadendorf,
D. (2014). BRAF, MEK and KIT inhibitors for melanoma:
Adverse events and their management. *Chinese Clinical
Oncology, 3*(3), 29. https://doi.org/10.3978/j.issn.2304-
3865.2014.03.03

Merlino, G., Herlyn, M., Fisher, D. E., Bastian, B. C.,
Flaherty, K. T., Davies, M. A., Wargo, J. A., Curiel-
Lewandrowski, C., Weber, M. J., Leachman, S. A.,
Soengas, M. S., McMahon, M., Harbour, J. W., Swetter,
S. M., Aplin, A. E., Atkins, M. B., Bosenberg, M. W.,
Dummer, R., Gershenwald, J., ... Ronai, Z. A. (2016).
The State of Melanoma: Challenges and Opportunities.
Pigment Cell & Melanoma Research, 29(4), 404–416.
https://doi.org/10.1111/pcmr.12475

Middleton, M. R., Grob, J. J., Aaronson, N., Fierlbeck,
G., Tilgen, W., Seiter, S., Gore, M., Aamdal, S., Cebon,
J., Coates, A., Dreno, B., Henz, M., Schadendorf, D.,
Kapp, A., Weiss, J., Fraass, U., Statkevich, P., Muller,
M., & Thatcher, N. (2000). Randomized phase III study
of temozolomide versus dacarbazine in the treatment
of patients with advanced metastatic malignant
melanoma. *Journal of Clinical Oncology: Official Journal of
the American Society of Clinical Oncology, 18*(1), 158–166.
https://doi.org/10.1200/JCO.2000.18.1.158

Miller, K. D., Siegel, R. L., Lin, C. C., Mariotto, A. B.,
Kramer, J. L., Julia H. Rowland, Kevin D. Stein, Rick

Alteri, & Ahmedin Jemal. (2016). *Cancer treatment and survivorship statistics, 2016.* 66(4). https://acsjournals. onlinelibrary.wiley.com/doi/full/10.3322/caac.21349 Niezgoda, A., Niezgoda, P., & Czajkowski, R. (2015). Novel Approaches to Treatment of Advanced Melanoma: A Review on Targeted Therapy and Immunotherapy. *BioMed Research International, 2015,* 851387. https://doi.org/10.1155/2015/851387

Palmer, S. R., Erickson, L. A., Ichetovkin, I., Knauer, D. J., & Markovic, S. N. (2011). Circulating serologic and molecular biomarkers in malignant melanoma. *Mayo Clinic Proceedings, 86*(10), 981–990. https://doi. org/10.4065/mcp.2011.0287

Rebecca, V. W., Sondak, V. K., & Smalley, K. S. M. (2012). A Brief History of Melanoma: From Mummies to Mutations. *Melanoma Research, 22*(2), 114–122. https:// doi.org/10.1097/CMR.0b013e328351fa4d

Serrone, L., Zeuli, M., Sega, F. M., & Cognetti, F. (2000). Dacarbazine-based chemotherapy for metastatic melanoma: Thirty-year experience overview. *Journal of Experimental & Clinical Cancer Research: CR, 19*(1), 21–34.

Soengas, M. S., & Lowe, S. W. (2003). Apoptosis and melanoma chemoresistance. *Oncogene, 22*(20), 3138–3151. https://doi.org/10.1038/sj.onc.1206454

Soffietti, R., Trevisan, E., & Rudà, R. (2012). Targeted therapy in brain metastasis. *Current Opinion in Oncology, 24*(6), 679–686. https://doi.org/10.1097/

CCO.0b013e3283571a1c

van Zeijl, M. C. T., van den Eertwegh, A. J., Haanen, J. B., & Wouters, M. W. J. M. (2017). (Neo)adjuvant systemic therapy for melanoma. *European Journal of Surgical Oncology: The Journal of the European Society of Surgical Oncology and the British Association of Surgical Oncology, 43*(3), 534–543. https://doi.org/10.1016/j.ejso.2016.07.001

Velho, T. R. (2012). Metastatic melanoma – a review of current and future drugs. *Drugs in Context, 2012.* https://doi.org/10.7573/dic.212242

Wilson, M. A., & Schuchter, L. M. (2016). Chemotherapy for Melanoma. In *Melanoma* (pp. 209–229). https://doi.org/10.1007/978-3-319-22539-5_3

Zarour, H. M., Tawbi, H., Tarhini, A. A., Wang, H., Sander, C., Rose, A., Fourcade, J. J., Chauvin, J.-M., Sun, Z., Pagliano, O., & Kirkwood, J. M. (2015). Study of anti-PD-1 antibody pembrolizumab and pegylated-interferon alfa-2b (Peg-IFN) for advanced melanoma. *Journal of Clinical Oncology, 33*(15_suppl), e20018–e20018. https://doi.org/10.1200/jco.2015.33.15_suppl.e20018

Chapter 9:

Ali, Z., Yousaf, N., & Larkin, J. (2013). Melanoma epidemiology, biology and prognosis. *EJC Supplements, 11*(2), 81–91. https://doi.org/10.1016/j.ejcsup.2013.07.012

American Cancer Society. (2019a, August 14). *What Is Melanoma Skin Cancer?* https://www.cancer.org/cancer/melanoma-skin-cancer/about/what-is-melanoma.html

American Cancer Society. (2019b, October 24). *Types of Cancers That Develop in Young Adults.* https://www.cancer.org/cancer/cancer-in-young-adults/cancers-in-young-adults.html

American Cancer Society. (2021, January 12). *Melanoma Skin Cancer Statistics.* https://www.cancer.org/cancer/melanoma-skin-cancer/about/key-statistics.html

Bataille, V. (2003). Genetic epidemiology of melanoma. *European Journal of Cancer (Oxford, England: 1990), 39*(10), 1341–1347. https://doi.org/10.1016/s0959-8049(03)00313-7

Bauer, J. (2010). Molecular epidemiology of melanoma. *Melanoma Research, 20,* e8. https://doi.org/10.1097/01.cmr.0000382758.93495.d6

Carr, S., Smith, C., & Wernberg, J. (2020). Epidemiology and Risk Factors of Melanoma. *Surgical Clinics of North America, 100*(1), 1–12. https://doi.org/10.1016/j.suc.2019.09.005

Centers for Disease Control and Prevention. (2012). *Principles of epidemiology in public health practice; an introduction to applied epidemiology and biostatistics. 3rd edition* (3rd Edition). U.S. Department of Health and Human Services. https://stacks.cdc.gov/view/cdc/6914

Chang, A. E., Karnell, L. H., & Menck, H. R. (1998). The National Cancer Data Base report on cutaneous and noncutaneous melanoma: A summary of 84,836 cases from the past decade. The American College of Surgeons Commission on Cancer and the American Cancer Society. *Cancer, 83*(8), 1664–1678. https://doi.org/10.1002/(sici)1097-0142(19981015)83:8<1664::aid-cncr23>3.0.co;2-g

Coggon, D., Rose, G., & Barker, D. (2003). *Chapter 1. What is epidemiology?* The BMJ | The BMJ: Leading General Medical Journal. Research. Education. Comment. https://www.bmj.com/about-bmj/resources-readers/publications/epidemiology-uninitiated/1-what-epidemiology

Garbe, C., & Leiter, U. (2009). Melanoma epidemiology and trends. *Clinics in Dermatology, 27*(1), 3–9. https://doi.org/10.1016/j.clindermatol.2008.09.001

Honardoost, M., Rajabpour, A., & Vakili, L. (2018). Molecular epidemiology; New but impressive. *Medical Journal of the Islamic Republic of Iran, 32*, 53. https://doi. org/10.14196/mjiri.32.53

Horrell, E. M. W., Wilson, K., & D'Orazio, J. A. (2015). Melanoma — Epidemiology, Risk Factors, and the Role of Adaptive Pigmentation. *In Melanoma — Current Clinical Management and Future Therapeutics.* IntechOpen. https://doi.org/10.5772/58994

Khoury, M. J. (1997, January 1). *Genetic Epidemiology.* https://www.cdc.gov/genomics/resources/books/ genepi2.htm

Koh, H. K., & Geller, A. C. (2011). The public health future of melanoma control. *Journal of the American Academy of Dermatology, 65*(5), S3.e1-S3.e4. https://doi. org/10.1016/j.jaad.2011.02.036

Marks, R. (2000). Epidemiology of melanoma. *Clinical and Experimental Dermatology, 25*(6), 459–463. https:// doi.org/10.1046/j.1365-2230.2000.00693.x

Matthews, N. H., Li, W.-Q., Qureshi, A. A., Weinstock, M. A., & Cho, E. (2017). Epidemiology of Melanoma. In W. H. Ward & J. M. Farma (Eds.), *Cutaneous Melanoma: Etiology and Therapy.* Codon Publications. http://www. ncbi.nlm.nih.gov/books/NBK481862/

MedlinePlus. (2020, August 18). *Xeroderma pigmentosum.* https://medlineplus.gov/genetics/condition/

xeroderma-pigmentosum/

National Institutes of Health. (2011, February 7). *How UV Radiation Triggers Melanoma*. National Institutes of Health (NIH). https://www.nih.gov/news-events/nih-research-matters/how-uv-radiation-triggers-melanoma

Nazario, A. L., Macheledt, J. E., & Vogel, V. G. (1995). *Epidemiology of Cancer and Prevention Strategies*. Cancer Network. https://www.cancernetwork.com/view/epidemiology-cancer-and-prevention-strategies

Oliveria, S. A., Christos, P. J., & Berwick, M. (1997). The role of epidemiology in cancer prevention. *Proceedings of the Society for Experimental Biology and Medicine. Society for Experimental Biology and Medicine* (New York, N.Y.), 216(2), 142–150. https://doi.org/10.3181/00379727-216-44164

Peto, J. (2001). Cancer epidemiology in the last century and the next decade. *Nature, 411*(6835), 390–395. https://doi.org/10.1038/35077256

Phillips, A. J. (1969). The Epidemiology of Cancer. *Canadian Family Physician, 15*(12), 44–47.

Rastrelli, M., Tropea, S., Rossi, C. R., & Alaibac, M. (2014). Melanoma: Epidemiology, Risk Factors, Pathogenesis, Diagnosis and Classification. *In Vivo, 28*(6), 1005–1011.

Sample, A., & He, Y.-Y. (2018). Mechanisms and prevention of UV-induced melanoma. *Photodermatology, Photoimmunology & Photomedicine, 34*(1), 13–24. https://doi.org/10.1111/phpp.12329

Siegel, R. L., Miller, K. D., & Jemal, A. (2020). Cancer statistics, 2020. CA: *A Cancer Journal for Clinicians, 70*(1), 7–30. https://doi.org/10.3322/caac.21590

Skin Cancer Foundation. (2021, January 13). Skin Cancer Facts & Statistics. *The Skin Cancer Foundation.* https://www.skincancer.org/skin-cancer-information/skin-cancer-facts/

Toporcov, T. N., & Filho, V. W. (2018). Epidemiological science and cancer control. *Clinics, 73*(Suppl 1). https://doi.org/10.6061/clinics/2018/e627s

Chapter 10:

"2020 Melanoma Skin Cancer Report: Steaming the Global Epidemic." (2020). Global Coalition for Melanoma Patient Advocacy. https://melanomapatients.org.au/wp-content/uploads/2020/04/2020-campaign-report-GC version-MPA_1.pdf

Bharath, A., & Turner, R. (2009). Impact of climate change on skin cancer. Journal of the Royal Society of Medicine, 102(6), 215–218. https://doi.org/10.1258/jrsm.2009.080261

Bridgman, A. C., Fitzmaurice, C., Dellavalle, R. P., Karimkhani Aksut, C., Grada, A., Naghavi, M., ... Drucker, A. M. (2020). Canadian Burden of Skin Disease From 1990 to 2017: Results From the Global Burden of Disease 2017 Study. Journal of Cutaneous Medicine and Surgery, 24(2), 161–173. https://doi.org/10.1177/1203475420902047

Cheung, W. Y., Bayliss, M. S., White, M. K., Stroupe, A., Lovley, A., King-Kallimanis, B. L., & Lasch, K. (2018). Humanistic burden of disease for patients with advanced melanoma in Canada. Supportive Care in Cancer, 26(6), 1985–1991. https://doi.org/10.1007/s00520-017-4025-9

Guy, G. P., Ekwueme, D. U., Tangka, F. K., & Richardson, L. C. (2012). Melanoma Treatment Costs. American Journal of Preventive Medicine, 43(5), 537–545. https://doi.org/10.1016/j.amepre.2012.07.031

Johnston, K., Levy, A. ., Lorigan, P., Maio, M., Lebbe, C., Middleton, M., … van Baardewijk, M. (2012). Economic impact of healthcare resource utilisation patterns among patients diagnosed with advanced melanoma in the United Kingdom, Italy, and France: Results from a retrospective, longitudinal survey (MELODY study). European Journal of Cancer (1990), 48(14), 2175–2182. https://doi.org/10.1016/j.ejca.2012.03.003

Melanoma. (2020) In Candian Dermatology Association. https://dermatology.ca/public-patients/skin/melanoma/

Memon, A., Bannister, P., Rogers, I., Sundin, J., Al-Ayadhy, B., James, P. W., & McNally, R. J. . (2021). Changing epidemiology and age-specific incidence of cutaneous malignant melanoma in England: An analysis of the national cancer registration data by age, gender and anatomical site, 1981–2018. The Lancet Regional Health - Europe, 2, 100024–. https://doi.org/10.1016/j.lanepe.2021.100024

Rees, M. ., Liao, H., Spillane, J., Speakman, D., McCormack, C., Donahoe, S., … Henderson, M.. (2016). Localized melanoma in older patients, the

impact of increasing age and
comorbid medical conditions. European Journal of
Surgical Oncology, 42(9), 1359–1366.
https://doi.org/10.1016/j.ejso.2016.01.010

The effect of socioeconomic status on risk of stage IV
melanoma. (2018). Journal of the
American Academy of Dermatology, 79(3), AB283–
AB283.
https://doi.org/10.1016/j.jaad.2018.05.1123

"World Population Ageing (highlights)." (2017). United
Nations. https://www.un.org/en/development/desa/
population/publications/pdf/ageing/WPA
017Highlights.pdf

Chapter 11:

Advances in Melanoma and Other Skin Cancers Research.
National Cancer Institute. (n.d.).
https://www.cancer.gov/types/skin/research.

*Castle Biosciences Announces Level of Evidence Review on
DecisionDx-Melanoma*
*Published in Latest Issue of American Journal of Clinical
Dermatology.* (2019) BioSpace.
https://www.biospace.com/article/releases/castle-
biosciences-announces-level-of-evidence-revie
W-on-decisiondx-melanoma-published-in-latest-issue-
of-american-journal-of-clinical-dermatology/.

Can Melanoma Be Found Early?: Finding Melanoma Early.
American Cancer Society. (n.d.).
https://www.cancer.org/cancer/melanoma-skin-
cancer/detection-diagnosis-staging/detection.html.

Ferris LK. Early Detection of Melanoma: Rethinking the
Outcomes That Matter. (2021)
JAMA Dermatol. doi:10.1001/jamadermatol.2020.5650.

Home. Skin Cancer | ABCDE Assessment for Melanoma
| Beaumont Health. (n.d.).
https://www.beaumont.org/conditions/melanoma/
abcde's-of-melanoma#:~:text=One%20easy%20
way%20to%20remember,when%20diagnosing%20
and%20classifying%20melanomas.

Levine, A., & Markowitz, O. (2018). Introduction to reflectance confocal microscopy and its use in clinical practice. *JAAD case reports*, 4(10), 1014–1023. https://doi.org/10.1016/j.jdcr.2018.09.019

Melanoma Skin Cancer Research: Melanoma Studies. American Cancer Society. (n.d.). https://www.cancer.org/cancer/melanoma-skin-cancer/about/new-research.html.

Nivolumab and Ipilimumab Effective against Melanoma That Has Spread to the Brain. National Cancer Institute. (2018). https://www.cancer.gov/news-events/cancer-currents-blog/2018/melanoma-brain-metastases-nivolumab-ipilimumab.

Sadozai, H., Gruber, T., Hunger, R. E., & Schenk, M. (2017). *Recent Successes and Future Directions in Immunotherapy of Cutaneous Melanoma.* Frontiers. https://www.frontiersin.org/articles/10.3389/fimmu.2017.01617/full#B1.

Shah , A., & VanderWalde, A. (2019). *Future Directions in Melanoma.* Future Directions in Melanoma Interview . https://www.practiceupdate.com/content/future-directions-in-melanoma/84835.

Valko-Rokytovská, M., Šimková, J., & Kostecká, M. M. andZ. (2017). *Possibilities for the Therapy of Melanoma: Current Knowledge and Future Directions.* IntechOpen. https://www.intechopen.com/books/human-skin-

cancers-pathways-mechanisms-targets-and-treat
ments/possibilities-for-the-therapy-of-melanoma-
current-knowledge-and-future-directions.

What Does Melanoma Look Like? National Cancer
Institute. (n.d.).
https://www.cancer.gov/types/skin/melanoma-
photos#:~:text=Melanoma%20is%20a%20type%
0of,the%20body%20with%20pigmented%20tissues.

Zhao B, He YY. (2010) Recent advances in the
prevention and treatment of skin cancer using
photodynamic therapy. *Expert Rev Anticancer
Ther;*10(11):1797-1809. doi:10.1586/era.10.154.

* 9 7 8 1 7 7 3 6 9 2 5 5 5 *